SpringerBriefs in Molecular Science

Chemistry of Foods

Series editor

Salvatore Parisi, Industrial Consultant, Palermo, Italy

More information about this series at http://www.springer.com/series/11853

Francesco Montanari · Veronika Jezsó
Carlo Donati

Risk Regulation in Non-Animal Food Imports

The European Union Approach

Francesco Montanari · Veronika Jezsó
Carlo Donati

 Springer

Francesco Montanari
Lisbon
Portugal

Veronika Jezsó
Budapest
Hungary

Carlo Donati
Rome
Italy

ISSN 2199-689X   ISSN 2199-7209 (electronic)
SpringerBriefs in Molecular Science
ISBN 978-3-319-14013-1   ISBN 978-3-319-14014-8 (eBook)
DOI 10.1007/978-3-319-14014-8

Library of Congress Control Number: 2014957703

Springer Cham Heidelberg New York Dordrecht London

Printed on acid-free paper

Springer International Publishing AG Switzerland is part of Springer Science+Business Media
(www.springer.com)

Preface

The idea of writing a book on the approach of the European Union to risk management in the area of import of feed and food of non-animal origin came to light during one session of the *Better Training for Safer Food* programme of the European Commission. In that context, in our capacity of tutors, we have had the opportunity to share our experience in the field with the staff of competent authorities of EU Member States, but also to learn directly from them about differences, difficulties and good practices that exist in this area across the EU.

The category of imports of non-animal origin includes both products of plant origin (e.g. fruits, vegetables, tubers, cereals, herbs, spices, sugar) and of mineral origin (e.g. salt). Rising consumer demands, in particular, for fresh produce during all the year makes the trade of such products increasingly global and considerable nowadays.

Traditionally subject to less stringent import requirements than those applying to products of animal origin, recent food scares triggered by fruits and vegetables— i.e. the *E. coli* in 2011 or the *norovirus* outbreak in 2012—have determined increased attention by policy-makers and scientists in Europe towards the health risks that those products may carry. At EU level, this approach has resulted in stricter import surveillance by national enforcement authorities on products of non-animal origin, on the one hand, and in the adoption of emergency measures, in certain instances. On the other hand, the same approach has led to the development and consequent implementation of a risk prioritisation exercise in this area that the European Food Safety Authority has been tasked with.

Against this background, this book first offers an overview of the main hazards that may be associated with imports of non-animal origin and their health implications (Chap. 1). This part is followed by a systematisation of the current EU regime applying to those products and an in-depth analysis of the specific risk-management measures currently in place (Chaps. 2 and 3).

Result of a multidisciplinary approach combining our scientific, public policy and legal expertise, this book is intended for academics, scientists, professionals, control authorities and for all those who desire to acquaint themselves with a very specific area of EU food law that is steadily acquiring its own standing. At the same time, we are aware that this area is continuously evolving because of the diversification of possible episodes and food scares with chemical causes and other hazards.

In spite of that, we hope that this work can help the reader to understand rationale, functioning and dynamics of the area of imports of non-animal origin and the correlated assessment of safety risks and chemical hazards in the EU.

Lisbon, October 2014   Francesco Montanari
Budapest   Veronika Jezsó
Rome   Carlo Donati

Contents

About the Authors

Francesco Montanari holds a Ph.D. in EU Law from the University of Bologna (Italy). He has over 14 years of professional experience mainly in the area of food safety, plant health and consumer protection. During his career, he worked several years for the retail sector in United Kingdom and at EU level on food labelling, nutrition and alcohol-related harm. He worked between 2009 and 2012 for the Directorate-General for Health and Consumers of the European Commission, where he was responsible for the enforcement of EU food safety standards at national level and for the development and the management of EU legislation in the area of import controls of feed and food of non-animal origin. Member of the European Food Law Association, he runs his own food practice in Lisbon and is Senior Associate at Arcadia International.

Veronika Jezsó holds an M.Sc. in Food Science from Corvinus University of Budapest (Hungary). She has more than 13-year experience in the area of food safety and consumer protection both in the private and the public sector. She has been an active member of numerous scientific and governmental committees both at local and international level. She worked between 2006 and 2012 for the Directorate General for Health and Consumers of the European Commission, where she was responsible for the development and the management of EU legislation on import control, in particular, in the area of feed and food of non-animal origin. Currently, she is an independent consultant on EU food safety and consumer policy.

Carlo Donati is a medical doctor from the University of Milan (Italy). He has over 20 years of experience in clinical research as a medical director in national and multi-national pharmaceutical companies. He served between 2003 and 2012 as an official of the Italian Ministry of Health with different responsibilities. He was first General Manager of the Italian 'Alliance against Cancer' network of cancer hospitals and participated in such a capacity in international delegations for the cooperation in health-care with other countries. Subsequently, he moved to the food safety area where he focused on imports of food of non-animal origin and development of national guidelines on good hygiene practices and HACCP principles. Currently, he is an independent consultant for risk assessment and quality assurance in the food sector.

Chapter 1
Risk Analysis, Contaminants and Impact on Health in Imports of Non-animal Origin: The EU Context

Carlo Donati

Abstract Following a general introduction on the main provisions of the Regulation (EC) No. 178/2002, the so-called General Food Law (GFL), this chapter considers the role played by risk analysis in underpinning the development of legislation in this area at the European Union (EU) level, including specific tasks of the European Food Safety Authority (EFSA) as risk assessor. The chapter also provides a close examination of the type of hazards and more frequently associated risks with imports of non-animal origin, including the impact on human health.

Keywords Contaminants · Food · Imports · Non-animal origin · Public health · Risk analysis

Abbreviations

3-MCPD	3-monochloropropane-1,2 diol ester
ALARA	As low as reasonably achievable
BfR	Bundesinstitut für Risikobewertung (Federal Institute for Risk Assessment)
b.w.	Bodyweight
EHEC	Entero-haemorrhagic *Escherichia coli*
EFSA	European Food Safety Authority
EU	European Union
FAO	Food and Agriculture Organisation
GFL	General Food Law
HACCP	Hazard Analysis and Critical Control Points
HUS	Haemolytic uraemic syndrome
DG SANCO	Directorate-General for Health and Consumers
JECFA	Joint FAO/WHO Expert Committee on Food Additives
MOE	Margin of exposure

F. Montanari et al., *Risk Regulation in Non-Animal Food Imports*, Chemistry of Foods, DOI 10.1007/978-3-319-14014-8_1

1

MRL   Maximum residue level
BIOHAZ   Panel on biological hazards
PCB   Polychlorinated biphenyls
PAH   Polycyclic aromatic hydrocarbons
RASFF   Rapid Alert System for Food and Feed
SCF   Scientific Committee on Food of the European Commission
WHO   World Health Organization
WTO   World Trade Organisation

1.1 Introduction

Following some major food scares in the 1990s, the European Union (EU) has identified food safety as one of its top priorities. First set out in the European Commission's White Paper on Food Safety of January 2000 (European Commission 2000), but still current today, the principle guiding EU action in this area is to apply an integrated approach ('from farm to fork') to the food chain, thus covering feed production, primary production, food processing, storage, transport and retail.

The Commission's White Paper set out the basis for a proactive food policy: the modernisation of the legislation into a coherent and increasingly transparent set of rules; the reinforcement of controls throughout all stages of the food chain; and the effective evidence of transparency, quality, excellence and independence of science underpinning EU food law, with the aim of achieving a high level of public health and consumer protection.

Basically, the key strategic priorities of the White Paper may be summarised as follows:

- The creation of the European Food Safety Authority (EFSA)
- The creation and the subsequent implementation of a consistent 'farm-to-fork' approach when developing the EU legislation in relation to feed and food safety
- The definition of the principle whereby feed and food business operators are primarily responsible for the safety of their products when placed on the market, whereas national authorities of EU Member States must ensure the appropriate verification of compliance by those operators with the applicable requirements.

These strategic priorities have been eventually adopted by means of the Regulation (EC) No. 178/2002, commonly known as 'General Food Law' (GFL) (European Parliament and Council 2002). Applicable as of 21 February 2002, the GFL lays down common principles for the development and the implementation of EU food law (Chapter II, Articles 4–10). In addition, the GFL also specifies the role and the obligations for feed and food business operators on the one hand and Member States on the other (Chapter II, Articles 14–21).

1.1.1 Objectives and Main Principles of the General Food Law

Overall, the GFL aims at ensuring a high level of protection of human life and health; the protection of animal health and welfare, plant health and the environment is also taken into account, as appropriate. At the same time, the free movement of safe food is an essential element of the EU internal market, besides contributing to the health and well-being of EU citizens as well as to their social and economic interests, in accordance with GFL, Article 5 paragraphs 1 and 2, and Recitals 1–3.

Indeed, EU citizens are entitled to safe food as well as to receive accurate and truthful information about the identity, features and the quality of purchased products (GFL, Articles 14 and 8). Because of the need of avoiding disparities among EU consumers as to the level of protection granted by national rules, the GFL makes sure that these latter are subject to a harmonised legal framework at the EU level, thereby guaranteeing that safety standards are generally the same in all EU Member States. Of course, this means also that the legal environment in which feed and food business operators trade does not differ significantly from one Member State to another. As a result, trade within the EU is facilitated.

Over the last decades, feed and food trade has become increasingly globalised. For this reason, the GFL provides that the EU must ensure that its safety requirements are developed in strict compliance with international standards. Indeed, the compliance with internationally agreed standards contributes significantly to overcoming trade barriers and, in doing so, helps deliver economic growth. Consequently, the EU plays an active role in the work of international standard-setting bodies, such as the Codex Alimentarius, the World Organisation for Animal Health and the International Plant Protection Convention. In addition, because of the necessity of guaranteeing fairness and consistency from an international trade perspective, the GFL requires that both imports and export activities into and from the EU have to comply with applicable requirements set by the EU food law (Articles 11 and 12).

As referred earlier above, the GFL also establishes some basic principles as regards responsibilities for feed and food safety. In particular, the primary responsibility for ensuring full compliance with EU food law requirements rests solely on feed and food business operators. In accordance with Article 17 paragraph 1 and Recitals 30 and 31, the same principle applies in relation to feed. On the other hand, the competent authorities of Member States must enforce EU food legislations by means of a system of official controls designed to verify that feed and food business operators fulfil their legal obligations, in accordance with Article 17 paragraph 2. In turn, the European Commission has to evaluate the effectiveness of the monitoring system maintained by the Member States through regular audits and inspections.

EU rules on official controls are based on the Regulation (EC) No. 882/2004 of the European Parliament and of the Council of 29 April 2004 (European Parliament and Council 2004a) on official controls performed to ensure the verification of compliance with feed and food law, animal health and animal welfare rules, as amended by successive regulations.

1.1.2 How Do EU Citizens Perceive Food-Related Risks?

After a first survey carried out in 2005, in 2010, EFSA run a second Eurobarometer poll in all EU Member States in order to assess how consumers' views on food-related risks had evolved over the past 5 years (TNS 2010). The survey was carried out on a representative sample of 26,691 individuals, aged 15 or older, who virtually represented the views of over 500 million of EU citizens. Some key findings of the survey are summarised in Table 1.1.

Table 1.1   Key findings of a Eurobarometer survey on food-related risks (2010)

Food-related risks and related perception in the EU
The majority of respondents associate food and eating with pleasure, such as selecting fresh and tasty foods (58 %) and with enjoyment of meals with friends and family (54 %). Food safety (37 %) is less commonly associated with food and eating as such
In the context of other potential risks, which are likely to affect them personally, the economic crisis (20 %) and environmental pollution (18 %) are viewed by more respondents as risks very likely to affect their lives than food-related problems (11 %). However, concern regarding food possibly damaging one's health has increased by 3 % points since 2005
19 % of citizens spontaneously cite chemicals, pesticides and other substances as the major concerns. This concern is confirmed by prompted responses: when offered a list of possible issues associated with food, 3 out of 10 Europeans mention chemical residues from pesticides (31 %), antibiotics (30 %) and pollutants such as mercury and dioxins (29 %), together with cloning animals for food products (30 %), as risks to be 'very worried' about
Citizens feel less confident in being able to personally deal with possible problems of chemical contamination (<40 %) and new technologies (<30 %)
When asked to indicate the extent to which they feel confident about various information sources, citizens express the highest levels of confidence in information obtained from health professionals and personal contacts: physician, doctor and other health professionals (84 % total confident) and family and friends (82 %). These information sources are closely followed by consumer organisations (76 %), scientists (73 %) and environmental protection groups (71 %)
National and European food safety agencies (EFSA) and European institutions attract a relatively high level of confidence at 64 and 57 %, respectively (national governments 47 %)
With respect to information on food safety matters, slightly more respondents appear to worry about the news they hear today compared to five years ago (26 % vs. 23 % in 2005), but fewer reports taking any action: in 2010, 11 % of citizens claim to have permanently changed their eating habits as a reaction to information on food safety (vs. 16 % in 2005)
The tendency to ignore information appears to be greater for information regarding diet and health (29 %) than for that concerning food safety (24 %)
There is broad agreement that public authorities do a lot to ensure that food is safe in Europe, that public authorities are quick to act, that they base their decisions on scientific evidence and that they do a good job in informing people about food-related risks. The level of agreement has increased compared to 2005 for all of these points
A majority of respondents believe that public authorities in the EU should do more (>80 % total agree) to ensure that food is healthy and to inform people about healthy diets and lifestyles. This view is consistent across all Member States

1.1.3   Principles of Risk Analysis

The EU has been at the forefront of the establishment of risk analysis as a methodological approach underpinning food policy orientations as well as of its international acceptance. Article 3 paragraph 10 of the GFL defines risk analysis as a process consisting of three interrelated components, i.e. risk assessment, risk management and risk communication, while Article 6 refers to it as the basis for the adoption of any food safety measure at the EU level.

Table 1.2 shows the main relevant definitions given by the GFL. In this context, 'hazard' is defined as *'a biological, chemical or physical agent in, or condition of, food or feed with the potential to cause an adverse health effect, while 'risk' means a function of the probability of an adverse health effect and the severity of that effect, consequential to a hazard'* (GFL, Article 3).

Clearly, not all requirements set by EU food law have a scientific basis: certain consumer information (e.g. origin labelling) or certain provisions designed to prevent misleading or fraudulent practices do not necessarily need a scientific foundation. However, whenever is needed, the scientific assessment of risk must be undertaken in an independent, objective and transparent manner, while relying on the best available science, in accordance with the GFL, Article 6 paragraph 2.

On the other hand, 'risk management' is the process of weighing policy alternatives in the light of results of the risk assessment and, if required, selecting the appropriate actions necessary to prevent, reduce or eliminate the risk so as to ensure the high level of health protection that is considered appropriate in the EU.

However, based on the joint reading of Article 6 paragraph 3 and Recital 19 in the GFL, EU decision-makers need to consider a number of potentially relevant elements, during the risk management phase, in addition to results of risk assessment. These elements include, for example, the feasibility of controls, but also social, ethical, traditional, economic and environmental considerations besides the

Table 1.2 Definitions of the three components of risk analysis as provided for in the GFL (Article 3 points 11–13)

Risk analysis in accordance with GFL: key definitions	
Risk assessment Article 3, point 11	'Means a scientifically based process consisting of four steps: hazard identification, hazard characterisation, exposure assessment and risk characterisation'
Risk management Article 3, point 12	'Means the process, distinct from risk assessment, of weighing policy alternatives in consultation with interested parties, considering risk assessment and other legitimate factors, and, if need be, selecting appropriate prevention and control options'
Risk communication Article 3, point 13	'Means the interactive exchange of information and opinions throughout the risk analysis process as regards hazards and risks, risk-related factors and risk perceptions, among risk assessors, risk managers, consumers, feed and food businesses, the academic community and other interested parties, including the explanation of risk assessment findings and the basis of risk management decisions'

precautionary principle (Sect. 1.2.2). These considerations are commonly known as 'other legitimate factors' and may, at times, justify risk management decisions deviating from what the outcome of risk assessment.

With reference to the other legitimate factors, it is interesting to note how the range of factors that may influence risk management decisions at EU level appears to be significantly broader in comparison with the viewpoint of the international law. Effectively, in the context of the Codex Alimentarius and of the World Trade Organisation (WTO), which takes the former as a reference for the development of international food safety and chemical standards, only 'health protection of consumers' and 'promotion of fair practices in food trade' may be taken into account by risk managers when risk assessments' results are being pondered (Codex Alimentarius Commission 2013). As far as WTO law is concerned, this strict interpretation is in line with the content of the Sanitary and Phytosanitary Agreement. This document, ignoring the difference between risk assessment and management in the context of risk analysis, merely stipulates that WTO members' relevant measures must be science based and never maintained without sufficient scientific evidence (Article 2.2).

The breakdown of risk analysis into risk assessment, risk management and risk communication that the GFL foresees is not purely theoretical, but has institutional relevance. Indeed, while the GFL tasks EFSA with the role of risk assessor within the EU risk analysis process (Article 22 paragraph 2), the role of risk manager is performed by the European Commission assisted as appropriate by EU Member States. As to risk communication, this area is a shared competence between EFSA and the European Commission (Article 40 paragraph 1 GFL). More precisely, while EFSA is responsible for risk communication in the fields of its mission, that is risk assessment, the Commission is competent for communication revolving around risk management decisions. Far from being of clear-cut understanding and, thus, often difficult to put in practice, the repartition of competences between EFSA and the Commission in the area of risk communication represents, possibly, the one and only grey area within the EU risk analysis process as a whole.

1.1.4 Transparency

Food safety and the protection of consumer interests are an increasing concern for governments, trade organisations, non-governmental organisations and the general public. For this reason, the GFL provides for greater stakeholders' involvement at all stages of the development, evaluation and revision of EU food legislations, thereby leading to increased consumer confidence in its standards and the way they are established (GFL, Article 9). Indeed, the consumer confidence is a key indicator of a successful food policy and is therefore a primary goal of the EU action in this area.

1.2 Risk Analysis: The Three Interrelated Components

As already anticipated, risk assessment is carried out independently from risk management in the EU food safety system. As a risk assessor at EU level, EFSA produces scientific opinions and provides technical advice. EFSA's advices constitute sound foundations for the development of EU policy and legislation in this area, while support the European Commission and EU Member States in taking the necessary risk management decisions.

EFSA's remit covers feed and food safety, nutrition, animal health and welfare, plant protection and plant health (Article 22 GFL). In all these fields, EFSA's mandate is to provide objective and independent advice based on the most up-to-date scientific information and knowledge.

One of the main features of EFSA's mission is the communication with the outside world; this approach is considered essential for improving food safety in the EU and building public confidence in the risk assessment process. For this reason, EFSA aims to issue effective, consistent, accurate and timely information, including scientific outputs, based on the risk assessments and opinions of its own Scientific Committee and Panels.

EFSA's Scientific Panels are responsible for EFSA's risk assessment work including the drafting of scientific opinions. Each Panel focuses on a different area of the feed and food chain, while the Scientific Committee has the task of supporting the work of the Panels on crosscutting issues and scientific matters of a horizontal nature. It focuses on developing harmonised risk assessment methodologies in fields where EU-wide approaches are not yet defined. The Scientific Committee and the Panels are composed of independent scientific experts with a thorough knowledge of risk assessment from universities, research institutions and national food safety authorities. All members are appointed through an open selection procedure on the basis of proven scientific excellence, including experience in risk assessment and peer-reviewing scientific work and publications (EFSA Scientific Committee and Panel Renewals 2014).

1.2.1 Risk Assessment

As shown in Table 1.2, risk assessment is a scientifically based process consisting of four steps: (i) hazard identification, (ii) hazard characterisation, (iii) exposure assessment and (iv) risk characterisation. The Procedural Manual of the FAO/WHO Codex Alimentarius Commission comprises the same definition (Codex Alimentarius Commission 2013).

The World Health Organization (WHO) defined risk assessment as the scientific evaluation of known or potential adverse health effects resulting from human exposure to foodborne hazards and provided the following additional information (WHO 1995):

- Hazard identification: The identification of known or potential health effects associated with a particular agent (i.e. types of injury and conditions of exposure).
- Hazard characterisation: The qualitative and/or quantitative evaluation of the nature of the adverse effects associated with biological, chemical and physical agents which may be present in food. For chemical agents, a dose–response assessment (analysis of the relationship between the dose received and the respective response) should be performed. For biological or physical agents, a dose–response assessment should be performed if the data are obtainable.
- Exposure assessment: The qualitative and/or quantitative evaluation of the degree of intake likely to occur.
- Risk characterisation: Integration of hazard identification, hazard characterisation and exposure assessment into an estimation of the adverse effects likely to occur in a given population, including attendant uncertainties. In other words, risk is the possibility that an adverse effect will occur.

The definition of risk assessment includes quantitative risk assessment, which emphasises reliance on numerical expressions of risk, and qualitative expressions of risk, as well as an indication of the attendant uncertainties.

Recently, the EFSA Scientific Committee has reviewed the use of risk assessment terminology within its Scientific Panel. As a result, the Scientific Committee has concluded that the risk assessment terminology may be considered as not completely harmonised within EFSA, after the examination of 219 opinions published by the Scientific Committee and Panels (EFSA Scientific Committee 2012).

1.2.2 Risk Management

According to the above-mentioned WHO report (1995), the four components of risk management frameworks can be summarised as follows:

- Preliminary risk management activities comprise the initial process. It includes the establishment of a risk profile to facilitate consideration of the issue within a particular context and provides as much information as possible to guide further action. As a result of this process, the risk manager may commission a risk assessment as an independent scientific process to inform decision-making.
- Evaluation of risk management options is the weighing of available options for managing a food safety issue in the light of scientific information on risks and other factors and may include reaching a decision on an appropriate level of consumer protection. Optimisation of food control measures in terms of their efficiency, effectiveness, technological feasibility and practicality at selected points throughout the food chain is an important goal. A cost–benefit analysis could be performed at this stage.

- Implementation of the risk management decision will usually involve regulatory food safety measures, which may include the use of Hazard Analysis and Critical Control Points (HACCP). Flexibility in the choice of individual measures applied by industry is a desirable element, as long as the overall programme can be objectively shown to achieve the stated goals. Ongoing verification of the application of food safety measures is essential.
- Monitoring and review is the gathering and analysing of data to give an overview of food safety and consumer health. Monitoring of contaminants in food and foodborne disease surveillance should identify new food safety problems as they emerge. Where there is evidence that required public health goals are not being achieved, rethinking of food safety measures will be needed.

In the activities related to risk management, the concept of 'precautionary principle' should be taken into account. A communication from the European Commission on the precautionary principle in 2000 (Communication 2000) observed that decision-makers are constantly faced with the dilemma of balancing the freedom and rights of individuals, industry and organisations with the need to reduce the risk of adverse effects to the environment, human, animal or plant health.

The precautionary principle should not be confused with the element of caution that scientists apply in their assessment of scientific data. Resorting to the precautionary principle presupposes that though potentially dangerous effects deriving from a phenomenon, product or process have been identified, scientific evaluation does not allow the risk to be determined with sufficient certainty. Decision-makers need to be aware of the degree of uncertainty attached to the results of the evaluation of the available scientific information. Establishing what is an acceptable level of risk for society is an eminently political responsibility.

The GFL portrays the precautionary principle as an option open to risk managers when decisions have to be made to protect health, but scientific information concerning the risk is, in some way, inconclusive or incomplete (Article 7). Therefore, in specific circumstances, provisional risk management measures necessary to ensure the high level of health protection chosen in the EU may be adopted, pending further scientific information enabling a more comprehensive risk assessment. These measures should be proportionate and no more trade restrictive than is required and should ultimately be reviewed within a reasonable period of time, depending on the nature of the risk to life or health identified and the type of scientific information needed.

1.2.3 Risk Communication

Risk communication is an integral and ongoing part of the risk analysis exercise, where ideally all stakeholder groups (such as industry and consumers) should be involved from the start (WHO 2009). Risk communication has the objective to make stakeholders aware of the process throughout risk analysis, i.e. when risk assessment is performed as well as when risk management decisions are taken.

Indeed, risk communication relating to food safety, as any other communication that is health related, requires prudent and in-depth consideration before being realised, since wrongful messages may have devastating effects on public opinion and on the economy.

From a risk assessment perspective, risk communication helps to ensure that the logic, outcomes, significance and limitations of such an assessment are understood and accepted by all the stakeholders. Furthermore, it may well occur that stakeholders hold information that may be of relevance for performing a full risk assessment: for instance, industry stakeholders may have unpublished data crucial to complement information risk assessors already have.

The matter has been recently discussed and redefined by EFSA with the issue of a detailed document (EFSA 2012a), because of the necessity of setting up a sort of practical and common framework in relation to risk communication activities.

1.3 Main Contaminants for Food of Non-animal Origin

1.3.1 General Food Safety Requirements

The GFL sets out general requirements whereby no unsafe food must be placed on the market. In detail, Article 14 of the GFL states circumstances and criteria for the plain definition of 'safe' food products (Table 1.3).

Table 1.3 Definition of 'unsafe food' according to the GFL, Article 14

Food unsafety in the EU according to the GFL
1. Food shall not be placed on the market if it is unsafe
2. Food shall be deemed to be unsafe if it is considered to be
• injurious to health
• unfit for human consumption
3. In determining whether any food is unsafe, regard shall be had:
• to the normal conditions of use of the food by the consumer and at each stage of production, processing and distribution and
• to the information provided to the consumer, including information on the label, or other information generally available to the consumer concerning the avoidance of specific adverse health effects from a particular food or category of foods
4. In determining whether any food is injurious to health, regard shall be had:
• not only to the probable immediate and/or short-term and/or long-term effects of that food on the health of a person consuming it, but also on subsequent generations;
• to the probable cumulative toxic effects;
• to the particular health sensitivities of a specific category of consumers where the food is intended for that category of consumers
5. In determining whether any food is unfit for human consumption, regard shall be had to whether the food is unacceptable for human consumption according to its intended use, for reasons of contamination, whether by extraneous matter or otherwise, or through putrefaction, deterioration or decay

1.3.2 Food Contaminants

Contaminants are substances that have not been intentionally added to food. These substances may be present in food as a result of one or more of the following steps: production, processing, packaging, transport and storage. Moreover, contaminants might result from environmental contamination.

According to Regulation (EEC) No. 315/93 of 8 February 1993, 'contaminant' means '*any substance not intentionally added to food which is present in such food as a result of the production (including operations carried out in crop husbandry, animal husbandry and veterinary medicine), manufacture, processing, preparation, treatment, packing, packaging, transport or holding of such food, or as a result of environmental contamination. Extraneous matter, such as, for example, insect fragments, animal hair, etc., is not covered by this definition*' (Council 1993).

At EU level, maximum levels for certain contaminants in food are currently set in Regulation (EC) No. 1881/2006 (European Commission 2006). Applicable as of 1 March 2007, the Regulation was subject to several amendments. Maximum levels in certain foods are set for the following contaminants: nitrate, mycotoxins (aflatoxins, ochratoxin A, patulin, deoxynivalenol, zearalenone, fumonisins), metals (lead, cadmium, mercury, inorganic tin), 3-monochloropropane-1,2 diol esters (3-MCPD), dioxins and dioxin-like polychlorinated biphenyls (PCB) and polycyclic aromatic hydrocarbons (PAH), in particular benzo[a]pyrene.

However, very often legislation identifies risks for public health, sets maximum allowed levels, gives instructions for sampling and analysis, but does not elaborate on the *reasons why* contaminants are dangerous for humans or for animals. Therefore, in an attempt at bridging the gap between legislators, feed and food business operators and enforcement authorities, this chapter aims at providing some basic information on the health effects of the main contaminants, thereby explaining *why* it is necessary to carry out controls in order to ensure that their levels are kept at the minimum level possible. Since the scope of this book is limited to products of non-animal origin, only the contaminants that are most frequently associated with them will be considered.

1.3.2.1 Acrylamide

Acrylamide is a chemical compound that forms in starchy food products during high-temperature cooking (including frying, baking and roasting). Acrylamide has been found in products such as potato crisps, French fries, bread, biscuits and coffee. From the chemical viewpoint, many studies have been carried out with the aim of understanding the mechanism of acrylamide formation in several foods. In summary, it can be affirmed that the production of acrylamide is strictly linked to the reaction of sugars with amino acids at high temperatures. In 2005, EFSA stated that acrylamide may be a human health concern, since it is considered to be both carcinogenic and genotoxic in animal tests, and that efforts should be made to reduce relevant exposure mainly through diet. EU Member States are requested to perform annual monitoring of acrylamide levels and feed them back to EFSA (European Commission 2010c). So far, this latter has assessed national data in the context of four annual monitoring reports. EFSA's last report (EFSA 2012b) did not reveal any considerable differences from previous years in the levels of acrylamide in most food categories assessed. More recently, Recommendation 2013/647/EU of the European Commission (2013a) provided new indicative values for acrylamide in a wider range of food matrices, including breakfast cereals, baby foods, coffee substitutes and processed cereal-based foods for infants and young children.

EFSA's Scientific Panel on Contaminants in the Food Chain (CONTAM Panel) is carrying out a full risk assessment of acrylamide in food. EFSA's experts provisionally completed this full risk assessment in July 2014. The Panel assessed the toxicity of acrylamide for humans and updated its estimate of consumer exposure through the diet. EFSA publicly consulted on their draft scientific opinion in mid-2014. EFSA will hold a follow-up meeting with stakeholders to discuss feedback received during the online consultation. The feedback from the public consultation will assist the Panel in finalising its scientific opinion, scheduled for the first half of 2015.

1.3.2.2 Dioxins and Dioxin-like Polychlorinated Biphenyls (PCB)

Dioxins and PCB are toxic chemicals that persist in the environment and tend to accumulate in the food chain. Although their presence in the environment in Europe has declined since the 1970s, following concerted efforts by public authorities and industry, they are still found, at low levels, in many food products. Long-term exposure to these substances has been proved to cause a range of adverse effects on the nervous, immune and endocrine systems in addition to impairing the reproductive function. PCB may also cause cancer. Their persistence, and the fact that they accumulate in the food chain, notably in animal fat, is thus still cause of some safety concerns.

At EU level, the European Commission has adopted Regulation (EU) No. 1259/2011, amending Regulation 1881/2006, to set maximum levels of dioxins in food (European Commission 2011a), and Regulation (EU) No. 277/2012 for feed (European Commission 2012a).

In 2012, EFSA published a new report on levels of PCB in food and feed. The report revealed a general decrease in dietary exposure to said contaminants, for the period 2008–2010 as opposed to 2002–2004 (decrease of, at least, 16 % and up to 79 % for the average adult population, with similar values for toddlers and other children). Exposure to non-dioxin-like PCB, a subset of PCB with different toxicological properties, also decreased (EFSA 2012c).

The new Commission Recommendation 2014/663/EU of 11 September 2014 has set new action levels for dioxins and dioxin-like PCB in feed and food (European Commission 2014).

1.3.2.3 Ethyl Carbamate

Ethyl carbamate is a compound that can naturally occur in fermented foods and beverages. It often occurs in alcoholic beverages (in particular stone fruit brandies). Ethyl carbamate is formed by ethanol and certain precursors in the fruit mash under the influence of light during the distillation process. Ethyl carbamate is genotoxic and a multisite carcinogen in animals and probably carcinogenic in human beings. On 2 March 2010, the European Commission adopted a recommendation on the prevention and reduction of ethyl carbamate contamination in stone fruit spirits and stone fruit marc spirits and on the monitoring of ethyl carbamate levels in these beverages (European Commission 2010a). The same document has also highlighted the necessity of a strict monitoring activity of this contaminant in the above-defined beverages (EFSA 2010).

1.3.2.4 3-Monochloropropane-1,2 Diol and Related Esters

3-MCPD is a food processing contaminant formed by heat as a reaction product of triacylglycerols, phospholipids or glycerol and hydrochloric acid in fat-based or fat-containing foods. Depending on the type of food, it may occur as a free substance, in the form of an ester with fatty acids or in both forms. With reference to adverse reactions and food safety issues, 3-MCPD toxicity seems to be mainly correlated with the kidney, with chronic oral exposure resulting in nephropathy and tubular hyperplasia and adenomas. Back in 2001, the Scientific Committee on Food of the European Commission (SCF) established a TDI for 3-MCPD of 2 µg/kg bodyweight (b.w.) (Scientific Committee 2001). Subsequently, the International Agency for Research on Cancer (IARC) has classified 3-MCPD as a possible human carcinogen (group 2B). The Joint FAO/WHO Expert Committee on Food Additives (JECFA) established for the free compound a provisional maximum tolerable daily intake (PMTDI) of 2 µg/kg b.w. More recently (2013), EFSA published a Scientific Report describing the levels of 3-MCPD in food, based on 1,235 analytical results collected from EU Member States from 2009 to 2011 (EFSA 2013a).

Moreover, in September 2014, the EU Commission published the Recommendation 2014/661/EU on the monitoring of the presence of 2- and 3-monochloropropane-1,2-diol (2- and 3-MCPD), 2- and 3-MCPD fatty acid esters and glycidyl fatty acid esters in food (European Commission 2014a).

At present, 3-MCPD has not been detected with appreciable frequency in foods. According to the EU Rapid Alert System for Food and Feed (RASFF), only one notification has concerned the presence of 3-MCPD in hydrolysed vegetable protein from Switzerland in the period 2009–2013.

1.3.2.5 Polycyclic Aromatic Hydrocarbons (PAH)

Polycyclic aromatic hydrocarbons (PAHs) constitute a large class of organic compounds that are composed of two or more fused aromatic rings. They are primarily formed by incomplete combustion or pyrolysis of organic matter and during various industrial processes. PAHs generally occur in complex mixtures which may consist of hundreds of compounds. Humans are exposed to PAHs by various pathways. While for non-smokers the major route of exposure is consumption of food, for smokers the contribution from smoking may be significant (Zanieri et al. 2007). Food can be contaminated from environmental sources, from industrial food processing and from certain home cooking practices.

Most of these compounds may be regarded as potentially genotoxic and carcinogenic to humans and therefore represent a priority group in the assessment of the risk of long-term adverse health effects following dietary intake of PAHs. At EU level, maximum levels for PAH, and in particular for benzo[a]pyrene, are currently laid down in Regulation 1881/2006. In 2008, EFSA published a scientific opinion on PAH in food (EFSA 2008).

At present, PAH has been detected with appreciable frequency in foods. According to the EU RASFF, 19 notifications have concerned the presence of benzo[a]pyrene in the period 2009–2013, mainly in vegetal oil and fish.

1.3.2.6 Metals

The contamination of foods and feeds is normally associated with organic chemicals such as PAH, dioxins and other substances. However, a notable and 'historical' tradition of food and non-food contamination concerns inorganic contaminants such as metals (arsenic, cadmium, lead, aluminium and mercury).

Metals can be naturally and widely found in the environment, e.g. soil, water and atmosphere, but the distribution of these elements may vary. In particular, food contamination by metals can have different (anthropic) causes: farming activities, industrial processes, emission of powders and other matters from industrial and car exhaust parts, food processing and storage. Exposure can thus take place through two distinct pathways: from the environment or by ingesting contaminated food or water. In addition, the danger is remarkably increased because of the known bioaccumulation of metallic elements in living tissues.

With reference to the EU legislation, maximum levels for metals such as cadmium, lead and mercury in certain foods have been established by Regulation (EC) No. 1881/2006.

Occurrence of unauthorised presence of aluminium has been frequently reported by EU Member States, e.g. in imported noodles or other pasta products. In this respect, it is worth recalling that in 2008, the EFSA Panel on Food Additives, Flavourings, Processing Aids and Food Contact Materials established a tolerable weekly intake (TWI) of 1 mg/kg b.w. from all sources of aluminium, whereas in 2011, the JECFA established a provisional tolerable weekly intake (PTWI) of 2 mg/kg b.w.

As for cadmium, a recent recommendation of the European Commission (April 2014) encourages EU Member States to ensure that farmers and food business operators steadily implement available mitigation measures for reduction of cadmium levels in food (in particular cereals, vegetables and potatoes) and monitor their progress (European Commission 2014b).

Over the period 2009–2013, the RASFF signalled some 150 notifications reported by Member States, due to the presence of metals in food. Some 20 notifications related to fruit and vegetables destined to import into the EU.

1.3.2.7 Mycotoxins

The category of 'mycotoxins' is one of the most important food contaminants in the modern industry because of the well-known toxic potential and the wide diffusion in foods.

A mycotoxin is a toxic secondary metabolite produced by organisms of the fungi kingdom, commonly known as moulds. The term 'mycotoxin' is usually reserved for the toxic chemical products produced by fungi that readily colonise crops. One mould species may produce many different mycotoxins, and the same mycotoxin may be produced by several species. On the other hand, aflatoxins are naturally occurring mycotoxins that are produced by *Aspergillus flavus* and *Aspergillus parasiticus*, species of fungi.

Under favourable environmental conditions, when temperature and moisture are conducive, these fungi proliferate and may produce mycotoxins. They commonly enter the food chain through contaminated food and feed crops, mainly cereals. The presence of mycotoxins in food and feed may affect human and animal health as they may cause many different adverse health effects, such as induction of cancer and mutagenicity as well as estrogenic, gastrointestinal and kidney disorders. Some mycotoxins are also immunosuppressive reducing resistance to infectious disease. In particular, aflatoxin B1 is a very potent carcinogen and a mutagen in many animals and is also a very potent human carcinogen.

In 1960, over 100,000 turkeys in British farms died in a few months from an apparently new disease that was called 'Turkey X disease'. Shortly after, however, also ducklings and pheasants were affected experiencing high mortality. The investigations conducted traced back the outbreaks to feed products, namely peanut meal with Brazilian origin. Further to the incident, speculations made regarding the nature of the toxin suggested that it might be of fungal origin. Indeed, the toxin-producing fungus was identified as *A. flavus* (1961) and the toxin was named aflatoxin after its origin (*A. flavus* → Afla) (Cornell University 2014). Inevitably, this discovery led to a growing awareness of the potential risks associated with these substances when present in feed and food.

From the technological and chemical viewpoints, aflatoxins pose a very serious problem because of the thermal stability. As a result, the effective reduction of contamination levels can be obtained by means of the physical separation of contaminated materials.

Actually, thermal treatments appear to reduce detectable amounts of mycotoxins in various foods, depending on the thermal level and the peculiar processing technique. However, the complete inactivation of mycotoxins cannot be assured. For this reason, several studies have also concerned the possibility of non-thermal technologies allowing mycotoxin destruction and inactivation. As an example, zearalenone and deoxynivalenol may be treated with pulsed light technology with interesting results (Moreau et al. 2013).

Maximum levels of aflatoxins (aflatoxins B1, B2, G1, G2 and M1) are laid down in Regulation 1881/2006 as amended by Regulation (EU) No. 165/2010 (European Commission 2010b) and by Regulation (EU) 1058/2012 (European Commission 2012b).

Special import conditions for products originating from certain non-EU countries presenting risk of contamination with aflatoxins are laid down in Regulation (EU) No. 884/2014 of 13 August 2014 (European Commission 2014c). Finally, several other imported products of non-animal origin (mainly peanuts and spices) are currently subject to the regime of reinforced border surveillance set by Regulation (EC) No. 669/2009 (European Commission 2009a).

Because of the complexity of the problem, a 'Guidance document for competent authorities for the control of compliance with EU legislation on aflatoxins' has been elaborated. This document is also applicable for the control of aflatoxins in food products not subject to the safeguard Regulation (Guidance 2010).

According to the latest RASFF report (RASFF 2014), the number of mycotoxin notifications decreased further significantly, which was due to a decrease in reported aflatoxin notifications. The decrease in aflatoxin notifications is mainly explained by the significant decrease in notifications related to the presence of aflatoxins in peanuts from India (from 88 notifications in 2012 to 15 in 2013) and in dried figs from Turkey (from 135 notifications in 2012 to 40 in 2013). On the other hand, in 2013, there were a significant number of notifications of aflatoxins in maize from the European region.

1.3.2.8 Pesticides

The category of 'pesticides' is a wide selection of contaminants with different functions and actions. Generally, the peculiar function may be of interest with reference to the classification. For this reason at least, the chemical classification of pesticides may be very difficult.

The term 'pesticides' covers insecticides, acaricides, herbicides, fungicides, plant growth regulators, rodenticides, biocides and veterinary medicines. Pesticides are chemical compounds used to

- Kill, repel or control pests to protect crops before and after harvest;
- Influence the life processes of plants;
- Destroy weeds or prevent their growth;
- Preserve plant products.

Because of the wide use of pesticides in the modern world and in the food sector, pesticides can be in contact with the human population in many ways, including food consumption.

Different negative health effects have been attributed to pesticides: neurological health effects (e.g. memory loss, loss of coordination, reduced speed of response to stimuli, reduced visual ability, altered or uncontrollable mood and general behaviour, and reduced motor skills), as well as asthma, allergies, hypersensitivity, cancer, hormone disruption, as well as problems with reproduction and foetal development.

With reference to the EU, the use of plant protection products and of their residues is carefully regulated.

In detail, Regulation (EC) No. 1107/2009 establishes uniform rules for their placing on the market (European Commission 2009b). On the other hand, Regulation (EC) No. 396/2005 (European Commission 2005a) establishes the maximum residue levels (MRLs) of pesticides permitted in products of plant or animal origin intended for human or animal consumption. According to this Regulation, MRL *'means the upper legal level of a concentration for a pesticide residue in or on food or feed set in accordance with this Regulation, based on good agricultural practice and the lowest consumer exposure necessary to protect vulnerable consumers'*. In other terms, defined MRLs are the result of a comprehensive assessment of the properties of the active substance and of the residue levels resulting from good agricultural practices defined for the treated crops. An indispensable precondition for setting MRLs is a risk assessment that demonstrates consumer safety (i.e. consumer intake not exceeding the toxicological reference values).

The same Regulation provides also the following definitions:

- Acute reference dose *'means the estimate of the amount of substance in food, expressed on a body weight basis, that can be ingested over a short period of time, usually during one day, without appreciable risk to the consumer on the basis of the data produced by appropriate studies and taking into account sensitive groups within the population (e.g. children and the unborn)'*;
- Acceptable daily intake *'means the estimate of the amount of substances in food expressed on a body weight basis, that can be ingested daily over a lifetime, without appreciable risk to any consumer on the basis of all known facts at the time of evaluation, taking into account sensitive groups within the population (e.g. children and the unborn)'*.

Over the last six years, EFSA has published several scientific opinions, in particular on cumulative risk assessment, which is the risk of a common toxic effect associated with concurrent exposure by all relevant pathways and routes of exposure to a group of chemicals that share a common mechanism of toxicity.[1]

---

[1] Examples of EFSA's Scientific Opinions include, but are not limited to (i) Scientific Opinion to evaluate the suitability of existing methodologies and, if appropriate, the identification of new approaches to assess cumulative and synergistic risks from pesticides to human health with a view to set MRLs for those pesticides in the frame of Regulation (EC) 396/2005, EFSA Panel on Plant Protection Products and their Residues (PPR), in EFSA Journal (2008) 704, pp. 1–85; (ii) Scientific Opinion on risk assessment for a selected group of pesticides from the triazole group to test possible methodologies to assess cumulative effects from exposure through food from these pesticides on human health, EFSA Panel on Plant Protection Products and their Residues (PPR), in EFSA J (2009) 7, 1167–1354; (iii) Scientific Opinion on the identification of pesticides to be included in cumulative assessment groups on the basis of their toxicological profile, EFSA Panel on Plant Protection Products and their Residues (PPR), in EFSA J (2013) 11, 3293–3424.

EFSA publishes an Annual Report on Pesticide Residues in the EU based on monitoring information of the official controls on pesticide residues in food received from all EU Member States and two EFTA countries (Iceland and Norway). The report assesses the exposure of European consumers to pesticide residues through their diets (EFSA 2013b).

In relation to importing activities, several imports of non-animal origin are currently subject to the EU regime of reinforced border surveillance for the possible occurrence of pesticide residues. On the other side, okra and curry leaves from India are currently subject to stringent import conditions set by Regulation (EU) No. 885/2014 because of the high level of non-compliant results (European Commission 2014d).

According to the RASFF Report (2014), concerning pesticide residues, in 2013—after a steady increase over several years—the amount of notifications for pesticide residues appears to have stabilised at 452, which is slightly more than the previous year. Only 36 of the notifications are about produce of EU origin.

1.3.3 Microbiological Contaminants

A large part of food safety concerns is directly correlated with the presence and the spreading of pathogen micro-organisms in food and feed products. Generally, most known and monitored foodborne diseases are associated with pathogens such as *Salmonella* sp., *Campylobacter jejuni*, entero-haemorrhagic *Escherichia Coli* (EHEC) and parasites such as *cryptosporidium*, *Cryptospora* and trematodes (WHO 2014).

In addition, the emergence of new pathogens and pathogens not previously associated with food is a major public health concern. At present, this phenomenon may be caused by modifications in farm practices, changes in food production and distribution, new food formulations and other factors.

The effective management of microbiological hazards can be truly enhanced using tools such as Microbiological Risk Assessment (MRA) and HACCP systems (WHO 2009).

MRA provides an understanding of the nature of the hazard: it can be considered a tool to set priorities for interventions. On the other hand, HACCP is a tool for process control through the identification of critical control points.

The United States' Centre for Disease Control and Prevention estimates that each year, only in that country, roughly 48 million people get sick, 128,000 are hospitalised, while 3,000 die because of foodborne diseases. As to the EU, in 2011, there were over 300,000 confirmed human cases of infection due to foodborne diseases, 15,000 hospitalisations and 300 deaths.

At the EU level, microbiological criteria for foodstuffs are laid down in Regulation (EC) No. 2073/2005 and its subsequent amendments (European Commission 2005b). This Regulation lays down the microbiological criteria for certain micro-organisms and the safety requirements to be complied with by food business operators when implementing the general and specific hygiene measures referred to in Article 4 of Regulation (EC) No. 852/2004 (European Parliament and Council 2004b).

The eradication or limitation of microbiological contaminants cannot be excluded when speaking of EU legislations in the field of food safety. However, this book is mainly focussed on chemical contaminants, including also toxins, while microbial agents and the correlated ecology in food and feed products should be discussed in other publications. For this reason, this chapter aims to highlight the importance of food safety concerns by microbiological causes without the detailed examination of historical situations. However, two of the most recent outbreaks of foodborne diseases caused by pathogens in imported food of non-animal origin may be briefly included here.

1.3.3.1 *Escherichia Coli* Outbreak (Mid-2011)

In May 2011, a cluster of three patients presenting symptoms of haemolytic uraemic syndrome (HUS) and hospitalised in the same health structure in Hamburg was reported. A Shiga toxin-producing *E. coli* (STEC) strain belonging to serotype O104:H4, a rare pathogen, was isolated from patients' specimens. Over the next few weeks, case counts increased rapidly in Germany and reached France and other countries, affecting persons who had recently travelled to Europe. Ill persons reported consuming locally grown raw sprouts. Subsequent investigations led by the European Commission and EFSA eventually identified one lot of fenugreek seeds imported from Egypt at the origin of the outbreak. On 6 July 2011, the EU temporarily banned importation of fenugreek and certain other seeds, beans and sprouts from Egypt (European Commission 2011b). Nonetheless, the final case count was of dramatic proportion with 4,075 cases ascertained (including 908 cases complicated by HUS) and 50 deaths in 16 different countries.

After the *E. coli* incident, the European Commission adopted the document *'Lessons learned from the 2011 outbreak of Shiga toxin-producing E. coli (STEC) O104:H4 in sprouted seeds'*, where areas for improvement and concrete actions were put forward (SANCO 2011). This, in turn, led to the introduction by the EU of microbiological criteria, including process hygiene criteria for sprout production and microbiological criteria for seeds for sprouting or for human consumption, in addition to permanent import requirements applying to imports destined to the EU market (i.e. Regulations (EU) No. 208/2013, 209/2013, 210/2013 and 211/2013) (European Commission 2013b–e).

1.3.3.2 Norovirus Outbreak (End 2012)

In autumn 2012, several cases of diarrhoea and vomiting occurred due to norovirus infections in children and adolescents who had eaten meals in schools and kindergartens in Germany. According to official statistics, the outbreak accounted for almost 11,000 cases, thus making it the biggest ever foodborne outbreak in that country. Investigations conducted by the German authorities pointed to a batch of deep-frozen strawberries of Chinese origin as the cause of the outbreak (BfR 2012).

Following the outbreak, Chinese strawberries from China were subjected to the EU system of reinforced border surveillance as of 1 January 2013 for possible contamination with norovirus and hepatitis A (at an increased frequency of testing of 5 %). Since controls were stepped up at EU borders, only two RASFF notifications (signalling presence of norovirus) in imports from China were issued. A recent audit of the Food and Veterinary Office (FVO) of the European Commission in China (DG SANCO 2013) identified a number of shortcomings in the system in place to control microbiological contamination in soft fruit intended for export to the EU. For instance, the audit revealed that apparently, no soft fruit processor considered viruses as a potential hazard under their HACCP plans. Also, despite increased surveillance by the Chinese competent authorities on products destined to export, the overall effectiveness of their official control system would be undermined by the inadequacy of the current sampling methods and laboratory capacity (DG SANCO 2013).

Following the food scares above referred, microbiological contamination in food of non-animal origin has been subject to increased monitoring and assessment. Some imports of non-animal origin (mainly herbs and spices) are currently subject to the EU reinforced border surveillance under Regulation 669/2009. Due to the high level of non-compliance detected at EU borders over the past three years, the EU has recently introduced a temporary ban on the import of paan leaves from Bangladesh for contamination with Salmonella (European Commission 2014e), amended by the Commission Implementing Decision 2014/510/EU (European Commission 2014f).

From its side, EFSA has recently come up with a scientific opinion on the risk posed by pathogens in certain food of non-animal origin (namely *Salmonella* spp. and norovirus in leafy greens) (EFSA BIOHAZ 2014). In its opinion, EFSA observes that each farm environment represents a unique combination of numerous characteristics that can influence the occurrence and the persistence of pathogens in leafy greens production. For this reason, EFSA considers that correct implementation of existing food safety management systems should be a primary objective for concerned farmers and producers.

1.4 Radionuclides

For a general viewpoint, the concept (and related fears) of radioactivity in the food chain is strictly linked to most recent nuclear accidents such as Chernobyl and Fukushima. Actually, there have not been signs of detectable radioactivity in most food categories until 2011. After this date, the Fukushima incident has raised public health concerns in Japan at least.

Basically, the internal effective dose of radionuclides in the human being can be increased depending on (a) the simple radon inhalation and (b) the consumption of contaminated food. Several factors correlated with foods can be important, including the type of food and the localisation of the production (D'Amato et al. 2013).

1.4.1 Chernobyl

On 26 April 1986, the Chernobyl nuclear power plant (Ukraine) suffered a major accident. As a result, a prolonged release of notable amounts of radioactive substances was observed in the atmosphere with the consequent widespread contamination of large areas in the European continent. According to recent studies that tried to determine current radiation exposure routes, wild foods (especially mushrooms and berries) and locally produced foods (e.g. milk and derived products) were found to act as main 'radioactive' carriers (Radiation Protection 2011).

Council Regulation (EC) No. 733/2008 currently sets accumulated maximum radioactive levels as to caesium-134 and caesium-137 of 370 Bq/kg for milk and milk products and certain products destined to infants and 600 Bq/kg for all other products concerned (European Council 2008). While these provisions apply to imports from non-EU countries, Commission Recommendation 2003/274/EURATOM also encourages EU Member States to take appropriate steps to ensure that the maximum permitted levels referred above are respected within the EU, when it comes to the placing on the market of wild game, wild berries, wild mushrooms and carnivorous lake fish (European Commission 2003).

EU efforts seem to have obtained good results so far. Over the period 2009–2013, EU RASFF signalled only eight cases where radioactivity levels were exceeded, mainly concerning mushrooms originating from East European countries.

1.4.2 Fukushima

The earthquake and tsunami that hit Japan on 11 March 2011 led to the release of radioactive material into the environment from Fukushima nuclear power station.

Alerted by the International Atomic Energy Agency (IAEA), the European Commission forthwith established an emergency team that operated on a 24-h/7-day basis for the following three weeks. In the wake of the accident, certain food products originating from the Japanese prefectures affected were reported to present radionuclide levels of concern. Since that contamination was considered a threat to public and animal health in the EU, the European Commission introduced emergency measures regulating imports of certain food products from Japan through the adoption of Regulation (EU) 297/2011 (European Commission 2011c). This latter was subject to several amendments in order to adjust the import conditions to the evolution of the situation in Japan. The Commission's latest update of the emergency measures in question is Regulation (EU) No. 322/2014, which was adopted in March 2014 (European Commission 2014g).

Since EU measures restricting imports from Japan have been in place, EU Member States did not notify any case of excessive radioactivity levels associated with Japanese exports through the RASFF. The absence of non-compliance cases can be reasonably attributed to the effectiveness of official controls performed by Japanese authorities prior to export.

Because of the known relationship between notable releases of radioactivity and the environment and potential acute and long-term health effects, the WHO has the duty of assessing public health risks with the consequent advice on measures to be taken. In relation to Fukushima, the relevant risk assessment was conducted by a pool of independent international experts selected by WHO for their expertise and experience in radiation risk modelling, epidemiology, dosimetry, radiation effects and public health. WHO health risk assessment concluded that no discernible increase in health risks from the Fukushima accident is expected outside Japan. As for the public health situation in Japan, the conclusions indicate that certain age and sex groups of the areas that were most affected by the incident may be subject to an increased risk of developing cancer during their lifetime. These findings, thus, still justify continued monitoring of food contamination and environmental impact (WHO/HSE/PHE 2013).

1.5 Emerging Risks

The identification of emerging risks in the field of feed and food safety is a key task assigned to EFSA, according to Articles 23 point (f) and 34 of the GFL. It should be highlighted that an emerging risk can be defined a risk resulting from a newly identified hazard to which a significant exposure may occur or from an unexpected new or increased significant exposure and/or susceptibility to a known hazard (EFSA 2007).

Against this background, EFSA is responsible for

- Collecting data and monitoring relevant information sources such as scientific literature, the RASFF and other international rapid alert systems, trade data and official bulletins.
- Developing procedures for analysing and evaluating collected data.
- Sharing information with stakeholders and Member States.

As a result, a yearly report is published by EFSA with relation to the overall strategy and specific activities on emerging risks. On the basis of the early identification of possible and potential risks in the food chain, EFSA can support risk managers in developing appropriate policy responses.

EFSA recently published a technical report that contains a systematic framework for the identification of emerging chemical risks occurring in the feed and food chain with a likely direct or indirect impact on human, animal and/or plant health. Such risks may arise from intentionally and non-intentionally produced industrial chemicals, as well as certain natural contaminants that may be transferred to the food or the feed chain through the environment (EFSA EN-547 2014).

1.6 The ALARA Principle and the MOE Approach

Potential risks in the food and feed chain are often correlated with the detection of genotoxic (i.e. damaging DNA, the genetic material of cells) and carcinogenic chemicals at the same time. At present, the detection of similar substances can be revealed at very low levels (i.e. less than 1 trillionth of a gram). For these reasons, risk managers have modified their approach in the last decade.

Until 2005, the best strategy for the reduction of exposure to potentially dangerous substances was the limitation of detected chemicals to 'a level that is as low as reasonably achievable' (ALARA principle). However, this approach has been considered inadequate because of the lack of bases for setting concrete priorities and actions, either with regard to the urgency or to the extent of measures that may be necessary (Alexander et al. 2012).

For these reasons, the EFSA Scientific Committee has proposed a new harmonised approach for the risk assessment of substances that are both genotoxic and carcinogenic (EFSA 2005): the margin of exposure approach (MOE). In detail, the addition or the use of substances that are both genotoxic and carcinogenic to foods or food components/ingredients in the food chain should not be allowed if residuals of these chemical compounds may be still found in foods.

In a document published in 2012, the EFSA Scientific Committee expressed the opinion that the MOE approach can be applied to impurities which are both genotoxic and carcinogenic, irrespective of their origin. The Scientific Committee reiterated its view expressed in 2005 that in general, a margin of exposure of 10,000 or higher would be of low concern from a public health point of view; the magnitude of an MOE, however, only indicates a level of concern and does not quantify risk. When using the MOE approach for assessing impurities, the derivation of the MOE, its magnitude and the uncertainties regarding its derivation should be described. A conclusion on whether the MOE is of high concern, low concern or unlikely to be of safety concern should also be provided (EFSA 2012d).

References

Alexander J, Benford D, Boobis A, Eskola M, Fink-Gremmels J, Fürst P, Heppner C, Schlatter J, van Leeuwen R (2012) Special issue: risk assessment of contaminants in food and feed. EFSA J 10(10):s1004–s1016. doi:10.2903/j.efsa.2012.s1004

BfR (2012) Tenacity (resistance) of noroviruses in strawberry compote. Opinion No. 038/2012, 06.10.2012. Bundesinstitut für Risikobewertung (Federal Institute for Risk Assessment), Berlin. Available: http://www.bfr.bund.de/cm/349/tenacity-resistance-of-noroviruses-in-strawberry-compote.pdf. Accessed 10 December 2014

Codex Alimentarius Commission (2013) Codex Alimentarius Commission: Procedural Manual, 21st edn. World Health Organization/Food and Agriculture Organization of the United Nations, Rome

Communication (2000) Communication from the commission on the precautionary principle. Brussels, 2.2.2000, COM (2000) 1 final. Available: http://ec.europa.eu/dgs/health_consumer/library/pub/pub07_en.pdf. Accessed 10 December 2014

Cornell University (2014) Aflatoxins: Occurrence and Health Risks. Cornell University, College of Agriculture and Life Sciences, New York. Available: http://www.ansci.cornell.edu/plants/toxicagents/aflatoxin/aflatoxin.html. Accessed 08 December 2014

Council (1993) Council Regulation (EEC) No. 315/93 of 8 February 1993 laying down Community procedures for contaminants in food. Off J L37:1–3

D'Amato M, Turrini A, Aureli F, Moracci G, Raggi A, Chiaravalle E, Mangiacotti M, Cenci T, Orletti R, Candela L, Di Sandro A, Cubadda F (2013) Dietary exposure to trace elements and radionuclides: the methodology of the Italian total diet study 2012–2014. Ann Ist Super Sanita 49(3):272–280. doi:10.4415/ANN_13_03_07

DG SANCO (2013) Final report of an audit carried out in China from 21 to 28 October 2013 in order to assess the control systems in place to control microbiological contamination in soft fruit intended for export to the European Union. DG(SANCO)/2013-6682 Final. Commission of the European Communities, Brussels

EFSA (2005) A harmonised approach for risk assessment of substances which are both genotoxic and carcinogenic. EFSA J 282:1–31

EFSA (2007) Definition and description of "emerging risks" within the EFSA's mandate (adopted by the Scientific Committee on 10 July 2007). Scientific Committee and Advisory Forum Unit, EFSA/SC/415 final. Available: http://www.efsa.europa.eu/en/scdocs/doc/escoem riskdefinition.pdf. Accessed 25 Aug 2014

EFSA (2008) Scientific opinion on polycyclic aromatic hydrocarbons in food, EFSA panel on contaminants in the food chain (CONTAM). EFSA J 724:1–114

EFSA (2010) Data collection templates for ethyl carbamate and 3-MCPD esters. EFSA J 8:1569–1574. doi:10.2903/j.efsa.2010.1569. Available: http://www.efsa.europa.eu/it/efsajournal/doc/1569.pdf. Accessed 10 December 2014

EFSA (2012a) When Food Is Cooking up a Storm. Proven Recipes for Risk Communications. European Food safety Authority, Parma. doi:10.2805/96864. Available: http://www.efsa.europa.eu/en/corporate/pub/riskcommguidelines.htm. Accessed 10 December 2014

EFSA (2012b) Update on acrylamide levels in food from monitoring years 2007 to 2010. EFSA J 10:2938–2976. doi:10.2903/j.efsa.2012.2938. Available: http://www.efsa.europa.eu/it/efsajournal/doc/2938.pdf. Accessed 10 December 2014

EFSA (2012c) Update of the monitoring of dioxins and PCBs levels in food and feed. EFSA J 10:2832–2914. doi:10.2903/j.efsa.2012.2832. Available: http://www.efsa.europa.eu/it/efsajournal/doc/2832.pdf. Accessed 10 December 2014

EFSA (2012d) Applicability of the margin of exposure approach for the safety assessment of impurities which are both genotoxic and carcinogenic in substances added to food/feed, Scientific Opinion of EFSA Scientific Committee. EFSA J 10(3):2578–2583. doi:10.2903/j.efsa.2012.2578

EFSA (2013a) Analysis of occurrence of 3-monochloropropane-1,2-diol (3-MCPD) in food in Europe in the years 2009–2011 and preliminary exposure assessment. EFSA J 11(9):3381–3426. doi:10.2903/j.efsa.2013.3381. Available: http://www.efsa.europa.eu/it/efsajournal/doc/3381.pdf. Accessed 10 December 2014

EFSA (2013b) The 2010 European Union Report on pesticide residues in food. EFSA J 11:3130–3938. doi:10.2903/j.efsa.2013.3130

EFSAEN-547 (2014) A systematic procedure for the identification of emerging chemical risks in the food and feed chain. EFSA (Supporting Publication) EN-547:1–40

EFSA BIOHAZ (2014) Scientific opinion on the risk posed by pathogens in food of non-animal origin. Part 2 (Salmonella and Norovirus in leafy greens eaten raw as salads). EFSA 12(3):3600–3718. doi:10.2903/j.efsa.2014.3600

EFSA Scientific Committee (2012) Scientific opinion on risk assessment terminology. EFSA J 10:2664–2707. doi:10.2903/j.efsa.2012.2664

EFSA Scientific Committee and Panel Renewals (2014) EFSA website. Available: http://www.efsa.europa.eu/. Accessed 17 Oct 2014

European Commission (2000) White paper on food safety. Brussels, 12.1.2000 COM (1999) 719 final. Available: http://ec.europa.eu/dgs/health_consumer/library/pub/pub06_en.pdf. Accessed 10 December 2014

European Commission (2003) Commission Recommendation 2003/274/EC of 14 April 2003 on the protection and information of the public with regard to exposure resulting from the continued radioactive caesium contamination of certain wild food products as a consequence of the accident at the Chernobyl nuclear power station. Off J Eur Union L99:55–56

European Commission (2005a) Regulation (EC) No. 396/2005 of the European Parliament and of the Council of 23 February 2005 on maximum residue levels of pesticides in or on food and feed of plant and animal origin and amending Council Directive 91/414/EEC. Off J Eur Union L70:1–16

European Commission (2005b) Commission Regulation (EC) No. 2073/2005 of 15 November 2005 on microbiological criteria for foodstuffs. Off J Eur Union L338:1–26

European Commission (2006) Commission Regulation (EC) No. 1881/2006 of 19 December 2006 setting maximum levels for certain contaminants in foodstuffs. Off J Eur Union L364:5–24

European Commission (2009a) Commission Regulation (EC) No. 669/2009 of 24 July 2009 implementing Regulation (EC) No. 882/2004 of the European Parliament and of the Council as regards the increased level of official controls on imports of certain feed and food of non-animal origin and amending Decision 2006/504/EC. Off J Eur Union L194:11–21

European Commission (2009b) Regulation (EC) No. 1107/2009 of the European Parliament and of the Council of 21 October 2009 concerning the placing of plant protection products on the market and repealing Council Directives 79/117/EEC and 91/414/EEC. Off J Eur Union L309:1–49

European Commission (2010a) Commission Recommendation 2010/133/EU of 2 March 2010 on the prevention and reduction of ethyl carbamate contamination in stone fruit spirits and stone fruit marc spirits and on the monitoring of ethyl carbamate levels in these beverages. Off J Eur Union L52:53–57

European Commission (2010b) Commission Regulation (EU) No. 165/2010 of 26 February 2010 amending Regulation (EC) No. 1881/2006 setting maximum levels for certain contaminants in foodstuffs as regards aflatoxins. Off J Eur Union L50:8–12

European Commission (2010c) Commission Recommendation 2010/307/EU of 2 June 2010 on the monitoring of acrylamide levels in food. Off J Eur Union L137:4–10

European Commission (2011a) Commission Regulation (EU) No. 1259/2011 of 2 December 2011 amending Regulation (EC) No. 1881/2006 as regards maximum levels for dioxins, dioxin-like PCBs and non dioxin-like PCBs in foodstuffs. Off J Eur Union L320:18–23

European Commission (2011b) Commission Implementing Decision 2011/402/EU of 6 July 2011 on emergency measures applicable to fenugreek seeds and certain seeds and beans imported from Egypt. Off J Eur Union L179:10–12

European Commission (2011c) Commission Implementing Regulation (EU) No. 297/2011 of 25 March 2011 imposing special conditions governing the import of feed and food originating in or consigned from Japan following the accident at the Fukushima nuclear power station. Off J Eur Union L80:5–8

European Commission (2012a) Commission Regulation (EU) No. 277/2012 of 28 March 2012 amending Annexes I and II to Directive 2002/32/EC of the European Parliament and of the Council as regards maximum levels and action thresholds for dioxins and polychlorinated biphenyls. Off J Eur Union L91:1–7

European Commission (2012b) Commission Regulation (EU) No. 1058/2012 of 12 November 2012 amending Regulation (EC) No. 1881/2006 as regards maximum levels for aflatoxins in dried figs. Off J Eur Union L313:14–15

European Commission (2012c) The Rapid Alert System for Food and Feed (RASFF) Annual Report 2011. Office for Official Publications of the European Communities, Brussels. Available: http://ec.europa.eu/food/safety/rasff/docs/rasff_annual_report_2011_en.pdf. Accessed 10 December 2014

European Commission (2013a) Commission Recommendation 2013/647/EU of 8 November 2013 on investigations into the levels of acrylamide in food. Off J Eur Union L301:15–17

European Commission (2013b) Commission Implementing Regulation (EU) No. 208/2013 of 11 March 2013 on traceability requirements for sprouts and seeds intended for the production of sprouts. Off J Eur Union L68:16–18

European Commission (2013c) Commission Regulation (EU) No. 209/2013 of 11 March 2013 amending Regulation (EC) No. 2073/2005 as regards microbiological criteria for sprouts and the sampling rules for poultry carcases and fresh poultry meat. Off J Eur Union L68:19–23

European Commission (2013d) Commission Regulation (EU) No. 210/2013 of 11 March 2013 on the approval of establishments producing sprouts pursuant to Regulation (EC) No. 852/2004 of the European Parliament and of the Council. Off J Eur Union L68:24–25

European Commission (2013e) Commission Regulation (EU) No. 211/2013 of 11 March 2013 on certification requirements for imports into the Union of sprouts and seeds intended for the production of sprouts. Off J Eur Union L68:26–29

European Commission (2014) Commission Recommendation 2014/663/EU of 11 September 2014 amending the Annex to Recommendation 2013/711/EU on the reduction of the presence of dioxins, furans and PCBs in feed and food. Off J Eur Union L272:17–18

European Commission (2014a) Commission Recommendation 2014/661/EU of 10 September 2014 on the monitoring of the presence of 2 and 3-monochloropropane-1,2-diol (2 and 3-MCPD), 2- and 3-MCPD fatty acid esters and glycidyl fatty acid esters in food

European Commission (2014b) Commission Recommendation 2014/193/EU of 4 April 2014 on the reduction of the presence of cadmium in foodstuffs. Off J Eur Union L104:80–81

European Commission (2014c) Commission Implementing Regulation (EU) No. 884/2014 of 13 August 2014 imposing special conditions governing the import of certain feed and food from certain third countries due to contamination risk by aflatoxins and repealing Regulation (EC) No. 1152/2009. Off J Eur Union L242:4–19

European Commission (2014d) Commission Implementing Regulation (EU) No. 885/2014 of 13 August 2014 laying down specific conditions applicable to the import of okra and curry leaves from India and repealing Implementing Regulation (EU) No. 91/2013. Off J Eur Union L242:20–26

European Commission (2014e) Commission Implementing Decision 2014/88/EU of 13 February 2014 suspending temporarily imports from Bangladesh of foodstuffs containing or consisting of betel leaves ('Piper betle'). Off J Eur Union L45:34–35

European Commission (2014f) Commission Implementing Decision 2014/510/EU of 29 July 2014 amending Implementing Decision 2014/88/EU suspending temporarily imports from Bangladesh of foodstuffs containing or consisting of betel leaves ('Piper betle') as regards its period of application. Off J Eur Union L228:33–34

European Commission (2014g) Regulation (EU) No. 322/2014 of 28 March 2014 imposing special conditions governing the import of feed and food originating in or consigned from Japan following the accident at the Fukushima nuclear power station. Off J Eur Union L95:1–11

European Council (2008) Council Regulation (EC) No. 733/2008 of 15 July 2008 on the conditions governing imports of agricultural products originating in Third Countries following the accident at the Chernobyl nuclear power station. Off J Eur Union L201:1–7

European Parliament and Council (2002) Regulation (EC) No. 178/2002 of the European Parliament and of the Council of 28 January 2002 laying down the general principles and requirements of food law, establishing the European Food Safety Authority and laying down procedures in matters of food safety. Off J Eur Comm L31:1–24

European Parliament and Council (2004a) Regulation (EC) No. 882/2004 of the European Parliament and of the Council of 29 April 2004 on official controls performed to ensure the verification of compliance with feed and food law, animal health and welfare rules. Off J Eur Union L165:1–141

European Parliament and Council (2004b) Regulation (EC) No. 852/2004 of the European Parliament and of the Council of 29 April 2004 on the hygiene of foodstuffs. Off J Eur Union L139:1–26

European Parliament and Council (2005) Regulation (EC) No. 396/2005 of the European Parliament and of the Council of 23 February 2005 on maximum residue levels of pesticides in or on food and feed of plant and animal origin and amending Council Directive 91/414/EEC. Off J Eur Union L70:1–16

Guidance (2010) Guidance document for Competent Authorities for the control of compliance with EU legislation on aflatoxins—November 2010. Available: http://ec.europa.eu/food/food/chemicalsafety/contaminants/guidance-2010.pdf. Accessed 10 December 2014

Moreau M, Lescure G, Agoulon A, Svinareff P, Orange N, Feuilloley M (2013) Application of the pulsed light technology to mycotoxin degradation and inactivation. J Appl Toxicol 33(5):357–363. doi:10.1002/jat.1749

Radiation Protection (2011) Radiation Protection No. 170: Recent scientific findings and publications on the health effects of Chernobyl. Working Party on Research Implications on Health and Safety Standards of the Article 31 Group of Experts, European Commission, Brussels

RASFF (2014) The Rapid Alert System for Food and Feed—2013 Annual Report. European Commission, Publications Office of the European Union, Luxembourg, 2014. ISBN 978-92-79-35992-7

SANCO (2011) Lessons learned from the 2011 outbreak of Shiga toxin-producing *Escherichia coli* (STEC) O104:H4 in sprouted seeds. Commission Staff Working Document, SANCO/13004/2011. Available: http://ec.europa.eu/food/food/biosafety/salmonella/docs/cswd_lessons_learned_en.pdf. Accessed 10 December 2014

Scientific Committee (2001) Opinion on 3-monochloro-propane-1,2-diol (3-MCPD), SCF/CS/CNTM/OTH/17 Final. European Commission, Health & Consumer Protection Directorate-General, Brussels. Available: http://ec.europa.eu/food/fs/sc/scf/out91_en.pdf. Accessed 10 December 2014

TNS (2010) Special Eurobarometer 354: Food-related risks, November 2010. TNS Opinion & Social, Brussels. Available: http://www.efsa.europa.eu/en/factsheet/docs/reporten.pdf. Accessed 10 December 2014

WHO (1995) Application of Risk Analysis to Food Standards Issues: Report of the Joint FAO/WHO Expert Consultation. WHO/FNU/FOS/Report No. 95.3. World Health Organization, Geneva

WHO (2009) Risk characterization of microbiological hazards in food. Microbiological Risk Assessment Series 17, Guidelines. World Health Organization, Geneva. Available. http://apps.who.int/iris/bitstream/10665/44224/1/9789241547895_eng.pdf?ua=1. Accessed 10 December 2014

WHO (2014) Food Safety. General Information Related to Microbiological Risks in Food. Available at http://www.who.int/foodsafety/areas_work/microbiological-risks/en/. Accessed 10 December 2014

WHO/HSE/PHE (2013) Health Risk Assessment from the nuclear accident after the 2011 Great East Japan Earthquake and Tsunami based on a preliminary dose estimation. World Health Organisation, WHO/HSE/PHE/2013 1(E):1–4

Zanieri L, Galvan P, Checchini L, Cincinelli A, Lepri L, DonzelliGP, Del Bubba M (2007) Polycyclic aromatic hydrocarbons (PAHs) in human milk from Italian women: influence of cigarette smoking and residential area. Chemosphere 67(7):1265–1274. doi:10.1016/j.chemosphere.2006.12.011

Chapter 2
Managing Risks in Imports of Non-animal Origin: The EU System of Reinforced Border Surveillance

Francesco Montanari

Abstract This chapter illustrates the full range of policy tools that are currently available to the European Union (EU) for counteracting risks associated with imports of feed and food of non-animal origin. While verification of compliance is performed by EU Member States by means of official controls, place and intensity of controls may vary depending on the seriousness of the risk to be addressed. Policy tools available to the European Commission and EU Member States in their capacity as risk managers include market surveillance, reinforced border controls, emergency measures, special import conditions and approval of checks prior to export. Following a general introduction on the European legal framework governing imports of non-animal origin, the present work analyses the main features of the EU system of reinforced border controls designed by the Regulation (EC) No. 669/2009.

Keywords Feed · Food · Imports · Non-animal origin · Non-EU countries · Official controls · Reinforced border surveillance

Abbreviations

BCP	Border Control Post
BIP	Border Inspection Post
CED	Common Entry Document
CHED	Common Health Entry Document
CP	Control Point
DPE	Designated Point of Entry
EFSA	European Food Safety Authority
EU	European Union

© The Author(s) 2015

F. Montanari et al., *Risk Regulation in Non-Animal Food Imports*, Chemistry of Foods, DOI 10.1007/978-3-319-14014-8_2

FVO   Food and Veterinary Office
GFL   General Food Law
ISO   International Standard Organisation
MANCP   Multi-Annual National Control Plan
OECD   Organisation for Economic Co-operation and Development
RASFF   Rapid Alert System for Food and Feed
SPS   Sanitary and Phytosanitary
TRACES   TRAde Control and Expert System
USA   United States of America
WTO   World Trade Organisation

2.1 Introduction

2.1.1 Official Controls: An Introduction

Following two major food scares during the 1990s, the European Commission published a 'White Paper on Food Safety' in January 2000 (European Commission 2000). Conceived as a blueprint document, the White Paper set down the overarching principles of the current legislative framework governing the food chain of the European Union (EU).

While calling for a more integrated approach in risk management along the food chain, the White Paper stressed the need to better clarify the obligations of the different stakeholders operating in said chain. It thus envisaged the setting up of a policy framework where feed and food business operators would be fully accountable for the safety of the products placed on the market and where public authorities would perform a supervising role.

Indeed, with regard to national authorities of EU Member States, subsequent EU legislation has clearly assigned them the task of ascertaining business operators' compliance with EU food safety, animal health or animal welfare requirements. In practice, this monitoring activity is possible by means of official controls.

Accordingly, Regulation (EC) No. 882/2004 (European Parliament and Council 2004) has introduced a set of harmonised provisions governing official controls performed by Member States' competent authorities, with effect as of 1 January 2006. Along with Regulation (EC) No. 178/2002 (European Parliament and Council 2002), commonly known as the 'General Food Law' (GFL), Regulation 882/2004 represents today a reference text for gaining a full understanding of this policy area.

2.1.2 Definition and Organisation of Official Controls

Regulation 882/2004 defines official controls as 'any form of control that the competent authority […] performs for the verification of compliance with feed and food law, animal health and animal welfare rules' (Article 2, point 1). In turn, 'verification', which is the ultimate purpose of official control, is intended as an activity encompassing the 'checking, by examination and the consideration of objective evidence, whether specified requirements have been fulfilled' (Article 2, point 2). EU Member States are responsible for the overall organisation of official controls and the efficacy of the relevant national control system. To this end, the Regulation specifies that official controls must be planned taking into account the risk involved and performed at an appropriate frequency and without prior warning (Article 3).

Furthermore, national competent authorities are responsible for ensuring that staff conducting official controls are adequately qualified and trained in their area of competence (Article 6). Generally, official controls are performed by either national central administrations or by any regional or local authority to which relevant powers have been conferred upon. On the other hand, specific tasks relevant to official controls may be delegated to independent control bodies, provided that certain conditions are met (Article 5).

2.1.3 Types of Checks

Usually, official controls on feed and food may consist of three different, though sequenced, types of checks: documentary, identity and physical checks. Table 2.1 portrays basic definitions for each of the checks in question.

Documentary checks may include the examination by official control authorities of commercial documents (e.g. invoices), transport documents (e.g. bill of lading, ship manifest for imports arriving at EU borders) or of any other document or certificate required by the EU law (e.g. veterinary, sanitary or phytosanitary certificates; results of laboratory analysis performed prior to export to the EU).

Physical checks may include activities such as sampling and laboratory testing. While sampling is normally conducted by an inspector pertaining to or, in

Table 2.1 The system of official controls in the EU according to Regulation (EC) No. 882/2004

Control activities: key definitions	
Documentary check Article 2 point 17	'Means the examination of commercial documents and, where appropriate, of documents required under feed or food law that are accompanying the consignment'
Identity check Article 2 point 18	'Means a visual inspection to ensure that certificate or other documents accompanying the consignment tally with the labelling and the content of the consignment'
Physical check Article 2 point 19	'Means a check on the feed or food itself which may include checks on the mean of transports, on the packaging, labelling and temperature, the sampling for analysis and laboratory testing and any other check necessary to verify compliance with feed or food law'

any event, under the supervision of the competent authority, analytical tests are carried out by official laboratories designated for this purpose by Member States and accredited against specific international standards such as EN ISO/IEC 17025 (Article 12, paragraphs 1 and 2).

2.1.4 Procedural Guarantees

Regulation 882/2004 provides feed and food business operators with a set of procedural rights when subject to official controls. For instance, when sampling for analysis is performed, the competent authority must comply with the following obligations (Article 11, paragraphs 5 and 6):

- Appropriate procedures must be in place to guarantee the right of the business operator to request a supplementary expert opinion, and, to this end,
- A sufficient number of samples must be made available to the same operator.

If one or more non-conformities are detected as a result of an official control, the competent authority would be also requested (Article 54, paragraph 3):

- To notify the concerned operator of the actions undertaken or those being planned in order to remove the non-compliance(s) identified, and
- To inform him/her of the reasons underpinning its decision as well as of his rights to appeal that decision.

2.1.5 Effectiveness of Official Controls

The European Commission verifies the effectiveness of the official control systems in the Member States by means of general and sector-specific audits carried out by the Food and Veterinary Office (FVO). While auditing a Member State, the FVO would normally verify (Article 45):

- The organisation and the functioning of the competent authorities responsible for the national control system
- The occurrence or persistence of problems in that context, as well as
- The level of implementation of the Multi-Annual National Control Plan (MANCP) that each Member State must have in place in accordance with Article 41 of Regulation 882/2004.

Moreover, the FVO carries out audits in non-EU countries, generally focussing on the assessment of the official control system applicable to feed and food products intended for export to the EU (Article 46).

With specific regard to the MANCP, it should be noted that Member States must submit annual reports on their implementation to the European Commission. The Commission is supposed to process the information received, together with relevant findings stemming from FVO audits or other sectoral reports, and produce an annual report on the overall operation of official controls in the EU (Article 44, paragraph 4).

The latest of these reports was published in October 2013, signalling a general improvement as regards the level of implementation of EU legislation on official controls (European Commission 2013a; González Vaqué 2013a). The report indicates that, overall, official controls organised by Member States appear to be increasingly risk-based, although not in same manner in all areas. In addition, available resources are often redeployed to ensure greater efficiency in the planning and execution of enforcement activities. The report claims, however, that there is some room for improvement especially as regards data collection: indeed, Member States do not compile (yet) relevant data in a way that is sufficiently consistent to cater for their full comparability.

2.1.6 Review of Regulation (EC) No. 882/2004

Regulation 882/2004 is currently under review. In May 2013, the European Commission launched a package of four legislative proposals in order to simplify and streamline the existing legislative framework governing animal health, plant health, seeds and official controls[1] (European Commission 2013b–e). The four proposals are currently under examination of the European Parliament and Member States in the Council.

In relation to the proposal on official controls (European Commission 2013c), the Commission intends to further strengthen the EU integrated approach to the food chain, by making rules on official controls applicable to areas that have been excluded so far (e.g. animal-by-products, veterinary residues) or only partially covered (e.g. plant health). Moreover, the proposal seeks to harmonise and improve EU rules concerning imports from non-EU countries (Sect. 2.2). Finally, since the proposal was finalised precisely when the horsemeat scandal hit the

---

[1] The four proposals have been announced as a part of the policy initiative 'Smarter rules for safer food' and preceded by the umbrella Communication from the Commission to the European Parliament and the Council with the title 'Healthier Animal and Plants and a Safer Agri-food chain', COM (2013) 264.

media in the EU,[2] it contains a number of new provisions aimed at improving EU and Member States' capability to detect food frauds and sanction them as appropriate (González Vaqué 2013b; Laurenza 2013; Montanari 2013).

At present, the outcome of the legislative negotiations is uncertain. It is however to be hoped that stakeholders involved will strive to agree on common solutions that will not lower or compromise the level of consumer protection that the EU has been able to ensure so far.

2.2 EU Import Control Policy

Provisions regarding official controls on import of animals, plants, feed and food into the EU are currently scattered across several EU legislative acts with horizontal or vertical scope and, sometimes, with a varying degree of harmonisation.

This piecemeal approach is mainly due to the different circumstances and policy drivers that, over time, have influenced the evolution of EU legislation relating to imports. In fact, the application of an integrated approach to the food chain, as introduced by the GFL and further enhanced through Regulation 882/2004, is a relatively recent milestone and, despite its merits, still incomplete.

Ultimately, the approach referred above has led to a complex framework consisting of different import control requirements and/or import conditions, depending on:

(a) The type of goods destined to the EU market (e.g. live animals, products of animal and non-animal origin, food contact materials, live plants and plant products), and/ or

(b) The seriousness of the risk involved.

---

[2] The horsemeat scandal can be regarded as the very first case of food fraud with an EU dimension. In essence, meat presented or labelled as beef was instead of *equidae* not destined to human consumption and thus used for being a cheaper ingredient. When control authorities of several Member States first detected the fraud in the first quarter of 2013, concerns arose that the fraud might also have some safety implications. Indeed, first sampling revealed, in certain instances, presence of residues of phenylbutazone, a veterinary drug usually administrated to horses employed in sports competition. For this reason, based on Article 53 of Regulation No. 882/2004, the European Commission recommended the organisation of a coordinated control plan across the EU. Results of the control plan indicated that, although the prevalence of veterinary residues was relatively limited, conversely, fraudulent mislabelling emerged as a widespread practice. In the wake of this fraud, the European Commission undertook a wide range of actions, including strengthening of requirements for controls on movement of horses within the EU, in order to prevent frauds of such an extent from happening again. Nearly one year after the scandal and while some criminal proceedings were still pending before national courts, in March 2014 the Commission launched a new round of coordinated controls aimed at ascertaining the prevalence of fraudulent practices as regards processing and labelling of bovine meat (Recommendation 2014/180/EU).

2.2.1 General Food Law and Imports

With regard to imports of feed and food, Article 11 GFL enshrines the fundamental principle whereby imported products, as much as EU products, must fulfil the relevant safety requirements laid down in EU law or 'equivalent conditions', where relevant. Yet, the EU may foresee, on emergency grounds, additional controls or conditions to be met, when faced with a serious risk associated with a product originating from outside the EU (Article 53 GFL). The margin of manoeuvre that the EU enjoys in such cases appears to be quite broad, with the spectrum of measures that can be adopted ranging from the suspension of relevant imports to any other interim measure that is deemed appropriate (including, for instance, the provision of export certificates and/or reinforced official controls on the EU territory).

2.2.2 Regulation (EC) No. 882/2004 and Import Controls

Regulation 882/2004 provides for a general framework for the organisation of official controls on imports of feed and food originating from non-EU countries (Chapter V, Articles 14 to 25).

Provisions of Regulation 882/2004 are supplemented by earlier Council Directives 91/496/EEC and 97/78/EC concerning, respectively, veterinary checks on imports of live animals and products of animal origin (Council of the European Communities 1991; Council of the European Union 1997). Although more than two decades have passed from the adoption of the above said Directives, the overarching principles in this area have remained substantially unchanged.

Essentially, import of veterinary goods into the EU is allowed only from authorised non-EU countries and, within their territory, from approved establishments. At their arrival in the EU, those goods must be presented at approved Border Inspection Posts (BIP) and accompanied by a veterinary certificate (to be duly signed by the competent authorities of the exporting country). While live animals entering the EU are subject to systematic documentary, identity and physical checks, feed and food of animal origin may be subject to reduced control frequencies for physical checks, if the relevant risk so allows.

In this respect, it is worth noting that the EU import control legislation in the veterinary sector is particularly strict as opposed to other areas. Indeed, as a rule, imports of feed and food of non-animal origin, including herbs, spices, fruits and vegetables, can freely enter the EU territory, unless there is a specifically regulated risk (Alemanno 2010).

As regards imports of feed and food of non-animal origin, the organisational principles set in Regulation 882/2004 are further elaborated by means of implementing legislation. As already anticipated, the overall philosophy governing this area is that specific EU import requirements may be set for cereals, fruits, vegetables intended for the EU market and verified at Designated Points of Entry (DPEs) whenever the risk they present so requires (Sect. 2.2.3).

Overall, one could question the reasons of maintaining two distinct regimes for imports of animal origin, on the one hand, and products of non-animal origin, on the other. In fact, several of the most recent food scares have been triggered from imported products of non-animal origin (Sects. 1.3.3.1 and 1.3.3.2). Recent events should at least call into question the current EU approach whereby products of non-animal origin are thought as involving a lower degree of chemical, physical or microbiological risk as opposed to products of animal origin. Finally, it is also worth recalling that the coexistence of different import control regimes may render the implementation of official controls a challenging task for national control authorities, besides being perceived as a potential hindrance to trade.

Despite the differences highlighted above, Regulation 882/2004 also foresees a set of provisions that are applicable to all imports, including products of non-animal origin. First among those provisions, Article 48 of this Regulation provides a legal basis for the adoption of specific import conditions for feed and food originating from non-EU countries, including lists of countries authorised to export specific products, export certificates and special import conditions. To some extent, the content of this provision may seem to overlap in part with Article 53 GFL. However, legislative practice have shown that Article 48 of Regulation 882/2004 has been applied in a relatively limited number of cases and, generally, with a view to setting import conditions on a permanent basis rather than tackling emergency situations that are, by definitions, limited in time.

Secondly, Article 49 of the same Regulation stipulates that the EU may recognise the equivalence of the official control system of another non-EU country with its own one, based on an international agreement or a favourable FVO audit. However, to date, the EU has not yet adopted an equivalence decision in accordance with this provision.

Thirdly, Article 23 of Regulation 882/2004 foresees the possibility for the EU to approve official controls carried out by an exporting country immediately prior to the dispatch of the relevant products to the EU. This approval may only be granted following appropriate verification that:

- The concerned feed and food exports are fully compliant with applicable EU requirements, and
- Official controls performed by the exporting country are conducted effectively enough to justify the replacement or the reduction of EU surveillance (Article 23, paragraph 3).

As a result of the granting of the approval, the intensity of import controls on the EU side is to be reduced. In any event, competent authorities in the Member States are required to maintain an adequate level of import surveillance in order to verify that the effectiveness of the pre-export checks approved by the EU remains unaltered. Similarly to Article 48, this provision has been used in a limited number of cases.

2.2.3 Towards a New EU Import Control Policy?

The European Commission seems conscious of the need of greater consistency across EU legislation on import controls. Indeed, in October 2010, the Commission published a 'Report on the effectiveness and consistency of sanitary and phytosanitary controls on imports of food, feed, animals and plants' (European Commission 2010d).

In essence, the Commission report acknowledges that further improvements may be necessary, in spite of the current system of EU import control functioning relatively well. One of the issues on which the report focuses is the need to make import controls and, in particular, physical checks more risk-based. Should this approach be embraced, the (limited) resources currently available at national level for official controls could be possibly better allocated and employed. In addition to that, the report notes that the differences across the various policy areas where import controls take place should be smoothed with a view to ensuring a more coherent EU import surveillance.

According to the Commission's proposal reviewing the framework for official controls (Sect. 2.1.6), in future all goods destined to import into the EU—be animals, plants, seeds food or feed—should be channelled through Border Control Posts (BCPs). The latter, located across the EU territory, should eventually replace the existing BIPs, DPEs and Points of Entry (for plants and plants products). In addition to that, arrival of consignments would require prior notification by business operators of a Common Health Entry Document (CHED). Depending on the technical feasibility, there are also plans to extend the use of the TRAde Control and Expert System (TRACES)—the electronic platform currently used in the veterinary sector to trace imports and intra-EU movement of live animals and products of animal origin—to other areas such as plants and plant products.

2.3 Imports of Feed and Food of Non-animal Origin: The Relevant EU Legal Framework

Article 15 of Regulation 882/2004 on official controls is at present the main provision governing EU policy on imports of feed and food of non-animal origin.

2.3.1 Market Surveillance

As a rule, feed and food products of non-animal origin from non-EU countries destined to import into the EU are subject to the control activities that Member States' national authorities perform in accordance with their MANCP. In this respect, Member States enjoy a relatively broad discretion in identifying those imports of non-animal origin that should be subject to official controls. In fact, Regulation 882/2004 merely prescribes that import controls in this area must be conducted on a regular basis, at an appropriate frequency and in light of potential risks (Article 15, paragraph 1). Similarly, the Regulation leaves Member States free to decide the most appropriate place for controls, including (Article 15, paragraph 2):

- The point of introduction into the EU territory
- The point of import
- Custom warehouses
- The premises of the importing business operator
- Any other stage within the feed or the food chain.

Generally, imports of non-animal origin that are merely subject to market surveillance do not raise serious safety concern and, from a risk management perspective, are regarded as low-risk products.

2.3.2 Reinforced Border Surveillance

Besides the general import regime set out in Article 15, paragraphs 1 and 2 of Regulation 882/2004 (Sect. 2.3.1), this provision foresees as well the establishment of a list of imports of feed and food of non-animal origin that, presenting a known or emerging risk, must be subject to reinforced surveillance at EU borders (Article 15, paragraph 5).

Overall, the EU reinforced surveillance regime for imports of non-animal origin is based on the following key elements (Regulation 882/2004, Article 17, paragraph 1):

(a) The designation of DPEs by EU Member States, i.e. ports, airports and terrestrial frontiers with non-EU countries equipped with adequate facilities
(b) The prior notification of the physical arrival of relevant consignments to the concerned DPE by business operators.

In accordance with Regulation 882/2004 (Articles 15, paragraph 5, and 17), the European Commission has adopted Regulation (EC) No. 669/2009 on an increased level of official controls on certain imports of feed and food of non-animal origin in July 2009[3] (European Commission 2009a). The Regulation formally entered into application on 25 January 2010.

---

[3] In this respect, it should be noted that the Regulation is based on Article 53 GFL as a legal basis in addition to Article 15 paragraph 5 of Regulation 882/2004. The use of Article 53 GFL as a legal basis can be explained considering that, when adopted, Regulation 669/2009 incorporated some imports that, at that time, were subject to emergency measures based on that provision (notably, Decision 2006/504/EC). This interpretation would be confirmed by the subsequent amendments of Annex I to Regulation 669/2009, which are systematically based only on Article 15 paragraph 5 of Regulation 882/2004.

The EU system of reinforced border surveillance introduced by Regulation 669/2009 has been designed in such a way that impact on trade should be minimal. Effectively, the system neither prevents the concerned products from entering the EU territory nor requires further assurances to be provided (e.g. sanitary certificates or analytical reports). Therefore, from a risk management perspective, imports of non-animal origin subject to reinforced border surveillance should be considered as presenting a 'medium risk'.

2.3.3   Emergency Measures

Whenever serious risks are associated with imports of non-animal origin, the latter may be subject to EU decisions commonly known as 'safeguard measures' or 'emergency measures'. These decisions may involve the suspension of imports or require compliance with other conditions that have trade-restrictive effects.

Generally, emergency measures are adopted to tackle situations that are considered serious from a public health perspective and, thus, based on Article 53 GFL. Two examples of emergency measures laying down specific requirements for imports of non-animal origin are as follows:

(a) Regulation (EU) No. 884/2014 (European Commission 2014e) imposing special conditions for the import of certain products from several non-EU countries because of aflatoxins contamination,[4]

(b) Regulation (EC) No. 258/2010 setting import requirements for guar gum from India (European Commission 2010a).

On the other hand, EU trade bans in this area are relatively rare and, generally, limited in time either because the emergency ceases to exist or following provision of appropriate guarantees by the exporting country. The EU decision to suspend imports of fenugreek seeds from Egypt following the outbreak of *E. coli* in Germany and France in 2011 (Sect. 1.3.3.1) is one recent example of a trade ban adopted at EU level in relation to imports of non-animal origin (European Commission 2011a, b). Most recently, the EU has introduced a temporary ban on imports of betel leaves from Bangladesh because of microbiological contamination by *salmonella spp* (European Commission 2014a, b).

Figure 2.1 exemplifies the existing relation between the seriousness of the risk that an import of non-animal origin may present (i.e. low, medium or high risk) and the policy instruments currently available to risk managers (i.e. European Commission together with Member States). Figure 2.1 also shows how:

(a) The seriousness of a given risk may increase from 'low' to 'medium' or 'high' and vice versa, depending on the circumstances of the case

(b) The qualification of a risk as 'low', 'medium' and 'high' ultimately impacts on the choice of the policy instrument to be used by risk managers.

---

[4] This Regulation replaces Regulation (EC) No. 1152/2009 (European Commission 2009b).

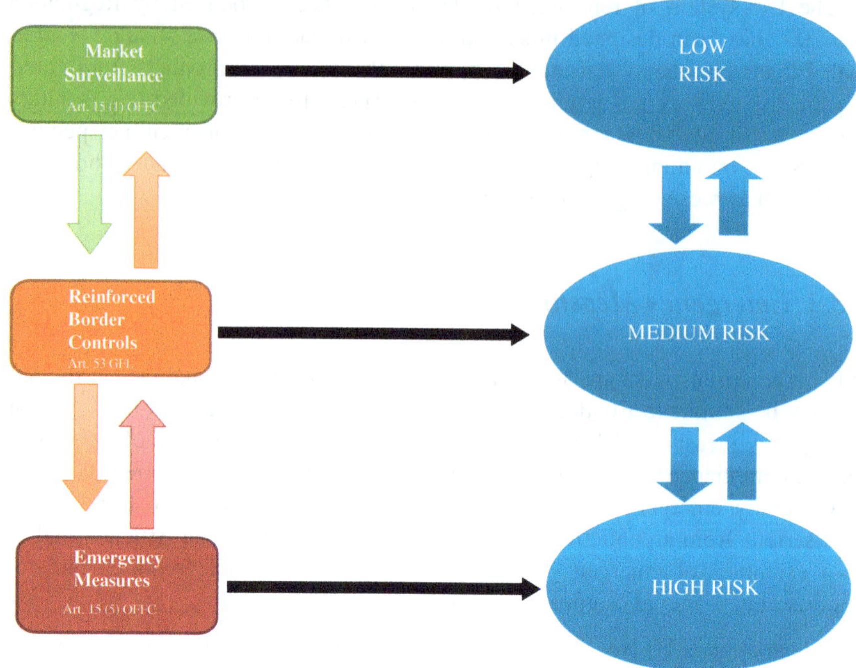

Fig. 2.1 Relation between risk and risk management policy tools

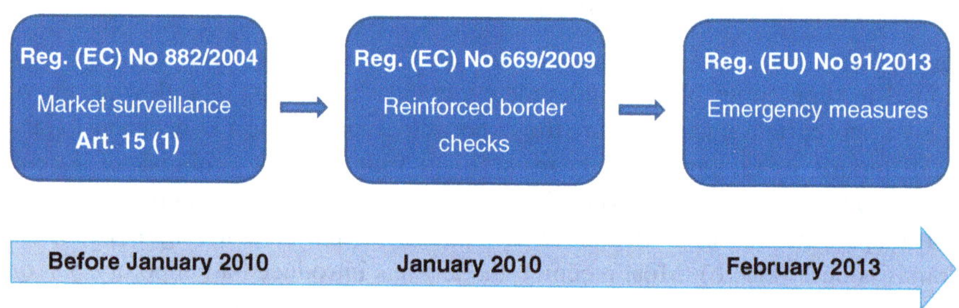

Fig. 2.2 Groundnuts and derived products from India for aflatoxin contamination. The EU legislation has progressively raised the attention level from the original 'low-risk' to 'high-risk' status

Two practical applications of EU risk management in this area are shown in Figs. 2.2 and 2.3.

2.3.4 Specific Import Conditions

As previously highlighted, Article 48 of Regulation 882/2004 is a provision of general relevance to all imports whereby specific import conditions (i.e. listing of

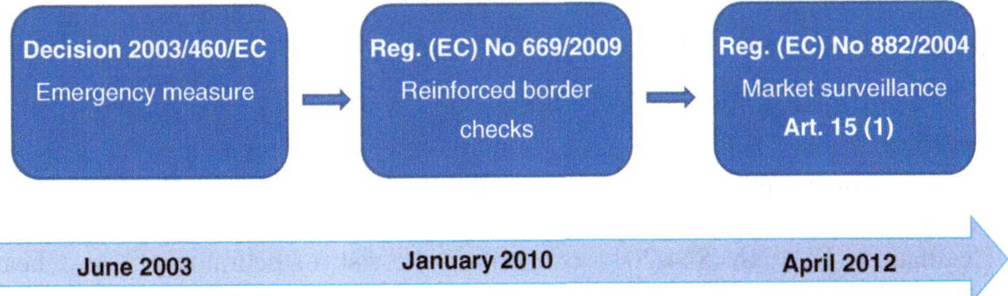

Fig. 2.3 *Chilli, chilli* products, curcuma and palm oil from all non-EU countries for Sudan dyes adulteration. The EU legislation has progressively lowered the attention level from the original 'high-risk' to 'low-risk' status

non-EU countries authorised to export, export certificates and/or special import conditions) can be imposed.

Since the entry into force of Regulation 882/2004, this provision has been used only twice. It was first used to adopt Regulation (EU) No. 284/2011 laying down specific import conditions on polyamide and melamine plastic kitchenware from China and Hong Kong (European Commission 2011c). More recently, the EU established specific certification requirements for sprouts and seeds for sprouting under Regulation (EU) No. 211/2013 (European Commission 2013f; Paganizza 2013; Rodriguez Font 2012). Adopted nearly 2 years after the biggest outbreak of *E. coli* the EU has ever known, this Regulation is part of a package of control measures, including, e.g. traceability requirements and listing of establishments approved for export, designed to prevent food-borne disease associated with such imports from happening again.

From the above, it would appear that the choice of Article 48 as a legal basis has to do with the permanent nature of the import conditions that the EU legislator can impose on the basis of that provision. On the other hand, an emergency situation, that is, by definition, limited in time, would be most appropriately addressed through measures based on Article 53 GFL.[5]

2.3.5 *Approval of Pre-export Checks*

As referred above (Sect. 2.2.2), Article 23 of Regulation 882/2004 provides the EU with a legal basis for the formal approval of the pre-export checks performed

---

[5] If the different rationale behind the two legal bases that can be used for imposing import requirements at EU level may appear clear in theory, practice shows that the EU legislator does not seem to abide by this interpretation systematically. For instance, one could question why Article 53 GFL is still used as a legal basis of Regulation 884/2014, when several imports covered (e.g. pistachios from Iran, groundnuts from Egypt) have been subject to import conditions for more than a decade. Hardly classifiable as emergency situations, those referred rather appear as recurring or structural safety issues, which, from a mere legal perspective, Article 48 Regulation 882/2004 would address more effectively.

by a non-EU country. Since the entry into force of the Regulation, use of this provision has been relatively limited. Indeed, it has been used only twice and namely in the area of imports of non-animal origin:

- Decision 2008/47/EC (European Commission 2008) regarding the risk of aflatoxins in peanuts and derived products from the United States of America (USA), and
- Regulation (EU) No. 844/2011 concerning the risk of ochratoxin A in wheat and wheat flour from Canada (European Commission 2011d).

Decision 2008/47/EC was adopted following a specific request submitted in 2005 by USA competent authorities as well as a FVO satisfactory audit. Overall, it allows the import peanuts and related products from USA provided that the following documents accompany the products:

- A health certificate contained in the Annex to the Decision with a 4-month validity from the date of issue,
- An analytical report containing results of sampling and analysis performed by an official laboratory of the exporting country in accordance with relevant EU standards.

With regard to official controls, in light of the guarantees provided by the exporting country, the Decision foresees that the frequency of physical checks should be 'significantly reduced' (Article 2). This means, that pre-export checks do not replace completely control activities at import stage. In other words, EU Member States must maintain a level of import surveillance that, though reduced, is proportionate to the risk (Regulation 882/2004, Article 16, paragraph 2). The notifications reported by Member States through the Rapid Alert System for Food and Feed (RASFF) over the past few years (2012: 5 detections, 2013: 6 detections) do confirm that national authorities still ensure a certain degree of surveillance on the products covered by the decision.

Similarly, Regulation 844/2011 requires the provision of a health certificate and an analytical report, both issued under the responsibility of the Canadian Grain Commission, attesting compliance of wheat and wheat flour from Canada, in particular, with EU standards for Ochratoxin A.

Concerning the reduced intensity for physical checks, the Regulation sets out a maximum 1 % control rate on all arriving consignments. Since 2011, official controls performed by EU Member States have apparently led to no detection.

2.4 Regulation (EC) No. 669/2009: Main Features of the System

Besides qualifying for as a special regime, the system of reinforced surveillance introduced by Regulation 669/2009 represents the first attempt, at EU level, at coordinating border control activities performed on imports of feed and food of non-animal origin by EU Member States.

Considering the impact of the novelties introduced by the Regulation, this latter has envisaged (Article 19):

- A six-month time interval between the entry into force of the Regulation (14 August 2009) and the actual date of its application (25 January 2010)
- A five-year transitional period in which control activities can be carried out away from the EU border (thus, from the premises of a DPE).

The provisions of Regulation 669/2009 have to be considered in conjunction with the Guidance document that the European Commission published following its entry into application.[6] Conceived as a 'living' document and aimed at ensuring uniform application of the control regime across the EU, the Guidance provides interpretations that, at times, go beyond the actual text of the Regulation. As an example, the document indicates the following categories of products as not being covered by the Regulation (and, thus, not subject to reinforced border surveillance):

- Composite products containing one or more of the feed or food listed in Annex I (e.g. salad or fruit mixes), unless otherwise specified by the Annex
- Products imported for research or laboratory purposes, as long as their intended use is well documented
- Products that are introduced in the territory of the EU for personal consumption
- Products transiting through the territory of the EU but destined to non-EU countries.

Further references to the content of the Guidance document will be made in the course of the following paragraphs as the main features of Regulation 669/2009 are presented.

2.4.1   Control Activities

As a rule, Regulation 669/2009 foresees that official controls on EU imports of non-animal origin requiring increased border surveillance should take place at the premises of DPEs. In terms of control activities, the Regulation requires relevant imports to undergo:

- Systematic (i.e. 100 %) documentary checks, and
- Identity and physical checks at a lower control rate (e.g. 5, 10, 20 and 50 %), as specified in Annex I.

---

[6] The European Commission's '*Question and Answers Paper on the provisions of Commission Regulation (EC) No. 669/2009 on an increased level of official controls on certain imports of feed and food of non-animal origin*' currently deals with 35 issues related to the application of that Regulation. Issues are grouped under six different headings: General concepts, Scope, Listing under Annex I, Implementation, Common Entry Document and, finally, Control Activities. The latest version of the Guidance document is available at the following web page: http://ec.europa.eu/food/food/controls/increased_checks/docs/QandA_paper_en.pdf.

Definitions of documentary, identity and physical checks contained in Article 2 Regulation 882/2004 (Table 1) apply also to control activities performed under Regulation 669/2009.

With regard to documentary checks, the Regulation sets out a precise timeframe (i.e. 2 working days) within which the competent authority of the DPE should perform them (Article 8 paragraph 1 (a)). On the other hand, the Regulation does not set a maximum timeframe for identity and physical checks: these controls should be carried out 'as soon as technically possible' (Article 8 paragraph 1 (b)). The lack of a specific timeframe may be justified if one considers the wide range of risks that Annex I to Regulation 669/2009 lists and the differences in turnaround times that may ensue from that.

In any event, feed and food business operators importing products covered by the Regulation 669/2009 may reasonably expect identity and physical checks (including delivery of the results of laboratory analysis) to be performed within the specific timeframes that other relevant EU legislation may set. Based on this interpretation, the 15-working day timeframe foreseen for the completion of official controls on products with high risk of aflatoxin contamination under Regulation 884/2014 would also apply to border controls on products listed for the same hazard in Annex I to Regulation 669/2009.

2.4.2 Designated Points of Entry

Designation of DPEs is an exclusive responsibility of EU Member States' competent authorities.[7] However, Regulation 669/2009 lists a set of minimum requirements that ports, airports or terrestrial borders must meet and namely (Article 4):

- Presence of suitably qualified and trained staff and in sufficient number
- Appropriate facilities for the performance of official controls including, where appropriate, storage and cold rooms, and a sheltered place where to perform sampling
- Appropriate unloading and sampling equipment
- Detailed instructions concerning sampling for analysis and the sending of samples to an accredited laboratory identified for this purpose
- An accredited laboratory, located at a reasonable distance from the DPE where samples can be sent for analysis.

Interestingly, the Regulation is silent as regards the possibility for veterinary BIPs and DPEs to share the same facilities. Should the use of shared facilities for veterinary and non-veterinary controls be allowed, national authorities of Member States would be reasonably expected to ensure that relevant hygiene standards are respected and cross-contamination is avoided.

---

[7] On the other hand, in the veterinary area, designation of BIPs is subject to the European Commission's prior approval and regular auditing by the FVO.

EU Member States must communicate to the European Commission the list of national DPEs. The Commission has, in turn, the obligation to make accessible the list of existing DPEs on its website.[8]

2.4.2.1 Use of Control Points

It should be noted that Regulation 669/2009 envisages a number of circumstances under which part of official controls may be performed away from DPEs. In any event, documentary checks must always be performed at the premises of the concerned DPE. Sections 2.4.2.1–2.4.2.3 illustrate the situations where derogations from the general border control regime are allowed.

As referred earlier above, Regulation 669/2009 has foreseen a five-year transitional period to provide EU Member States with sufficient time to equip DPEs (Article 19). However, due to practical difficulties encountered by some Member States in establishing DPEs within the timeframe initially foreseen, Regulation (EU) No. 718/2014 has prolonged the transitional period until 14 August 2019 (European Commission 2014c).

During the transitional period, Member States may decide to make use of control points (CPs) located away from EU borders for the performance of identity and physical checks. In any event, CPs must fulfil the same minimum requirements foreseen for DPEs and be authorised by the competent authorities of the concerned Member State. Article 19, paragraph 2 clarifies that Member States have to make available to the public relevant details of CPs: as a result, CPs are listed together with DPEs in national lists as required by Article 5 of Regulation 669/2009.

2.4.2.2 Special Geographical Circumstances

Article 9, paragraph 1, of Regulation 669/2009 provides for the possibility of performing physical checks at the premises of the feed or food business operator, whenever the DPE operates under special geographical circumstances, i.e. location in mountainous areas, ports or airports located on a small island. EU Member States interested in obtaining this derogation must seek an authorisation with the European Commission.

At present, only two Member States have requested such an authorisation. The following DPEs are authorised to perform part of the controls under Regulation 669/2009 at the premises of approved business operators (European Commission 2010b, c):

- Floriana port (Malta)
- Larnaca airport (Cyprus)
- Limassol port (Cyprus).

---

[8] The full list of DPE is available at the following web page: http://ec.europa.eu/food/food/controls/increased_checks/list_DPE_en.htm.

The relevant Commission's decisions clarify that the authorisations are granted, in principle, on a permanent basis, unless the assurances provided by the concerned Member State cease to exist.

2.4.2.3 Highly Perishable Products and Packaging with Special Characteristics

Regulation 669/2009 also foresees (Article 9, paragraph 2) that, under exceptional circumstances, identity and physical checks may be carried out directly by the competent authorities of the place of destination of the consignment, i.e. anywhere in the EU territory. These arrangements would be justifiable, in particular, when sampling at DPE might compromise the safety or the quality of products to an unacceptable extent. More precisely, the concerned products should be 'highly perishable'; alternatively, their packaging materials should present special characteristics.

Whether a product is 'highly perishable' or its packaging has special characteristics, these are requirements that have to be specified in Annex I to Regulation 669/2009. At present, no import of feed or food of non-animal origin has been listed in Annex I under such conditions, although the possibility of making use of the regime under discussion has been considered a few times (e.g. in relation to herbs and spices or hazelnut paste in vacuum packages).

2.4.2.4 Onward Transportation

Article 8, paragraph 2, of Regulation 669/2009 foresees that competent authorities of a DPE may authorise the onward transportation of the products to be imported until their final destination, while results of physical checks are still pending. Competent authorities of the DPE may consider taking this decision, in particular, when perishable products such as fruits and vegetables are subject to border controls. Current practice shows that several Member States allow onward transportation.

The purpose of this provision is quite evident: by allowing onward transportation, the EU legislator wants to ensure that the impact of Regulation 669/2009 on trade is minimised.

Some difficulties may arise when the final destination of the consignment is in the territory of a Member State other than the one through which the products have entered the EU. In such cases, the coordination and the communication between competent authorities of the DPE, on the one hand, and the competent authorities of the place of destination, on the other, are not always as smooth as they should.

For this reason, besides providing further clarifications in this area through its Guidance, Q&A n. 16 (Sect. 2.4), the European Commission has set up a network of competent authorities in Member States with a view to ensuring appropriate implementation of onward transportation throughout the EU.

2.4.2.5 Outcome of Official Controls

Release for free circulation into the EU of consignments under Regulation 669/2009 is possible only:

- Upon presentation of a Common Entry Document (CED) to the competent custom office by the feed or food business operator (Sect. 2.5)
- Whether all official controls (i.e. documentary checks and, where applicable, identity and physical checks) have been carried out successfully.

In this regard, Regulation 669/2009 fully mirrors the approach followed by the EU legislator for other emergency measures applying to imports of non-animal origin (European Commission 2009b, 2013g, 2014e, f). The same can be said for the provision of the Regulation concerning splitting of consignments (Article 12): splitting is allowed only after the completion of official controls by the competent authorities.[9]

Official controls performed pursuant to Regulation 669/2009 may lead to the identification of one or more non-compliances. The latter may consist of the following:

(a) Failure of documentary checks (e.g. absence or inadequate completion of CED)
(b) Unfavourable results of laboratory tests, whenever physical checks are required.

For such cases, Article 13 of Regulation 669/2009 refers back to the set of actions that competent authorities may undertake following detection of non-compliant consignments in accordance with Articles 19, 20 and 21 of Regulation 882/2004. These actions include destruction, special treatment and redispatch.

2.5 Obligations for Feed and Food Business Operators

Regulation 669/2009 lays down some requirements for feed and food business operators willing to import into the EU products that fall into its scope.

First of all, business operators must present their feed or food products to a DPE and ensure adequate prior notification—one working day in advance—of the physical arrival of the consignment (Article 6). The prior notification involves the submission (by fax, e-mail or through electronic platforms such as TRACES) of a CED (Annex II to the Regulation) and, in particular, completion by the business operator (or its representative) of Part I of that document.[10] It is worth noting that

---

[9] Furthermore, all these provisions foresee that, following splitting, each part of the consignment must be accompanied by an authenticated copy of the original CED until it is released for free circulation.

[10] On the other hand, Parts II and III of the CED are to be filled in by the competent authorities.

at present several Member States are using TRACES, on a voluntary basis, for tracing and recording information on imports of non-animal origin (Sect. 2.2.3).

Secondly, business operators importing consignments with special characteristics are expected to provide staff at the DPE with the unloading and sampling equipment that may be necessary for the performance of official controls.

Thirdly, business operators are required to bear costs of official controls: these costs have to be paid in the form of fees (Regulation 669/2009, Article 14, paragraph 2). For the determination of fees occasioned by reinforced border controls, Regulation 669/2009 refers to the relevant provisions of Regulation 882/2004. This approach results in fees not being fully harmonised across in the EU and, in certain instances, with considerable differences between Member States.

Eventually, it is worth mentioning that business operators whose products undergo reinforced controls under Regulation 669/2009 are also subject to general obligations and rights set in EU food law. They are thus required to ensure the safety of feed and food products they import and market into the EU territory in accordance with the GFL, besides being accountable for any civil and/or criminal liability directly stemming from breaches of EU feed and food safety requirements (Regulation 882/2004, Article 1, paragraph 4). On the other hand, whenever subject to official controls, they enjoy the procedural guarantees provided by Regulation 882/2004 (Sect. 2.1.4).

2.6   Obligations for Competent Authorities in the Member States

Regulation 669/2009 lays down a number of specific requirements also for national authorities of EU Member States. As referred above, Member States are responsible for the designation of DPEs and must have system in place to levy fees occasioned by official controls. In addition to that, Member States must regularly inform the European Commission of the results of their border control activities. More precisely, they must submit quarterly reports detailing the following information (Article 15):

- Number of incoming consignments of imports listed under Annex I to Regulation 669/2009 and relevant volumes,
- Number of consignments sampled and analysed,
- Number of non-compliances detected.

Based on this information, the European Commission and EU Member States can ensure a continuous assessment of the food safety risks that are associated with the imports listed in Annex I.

Finally, the general obligations applying to Member States as per Regulation 882/2004 are also relevant in the context of the reinforced controls required by Regulation 669/2009. One of these obligations is, for example, the requirement to have in place, at national level, proportionate, dissuasive and effective penalties for sanctioning non-compliances (Regulation 882/2004, Article 55).

2.7 Audits of the Food and Veterinary Office

During the period 2010–2011, the FVO has carried out a series of audits to evaluate official controls on imports of food of non-animal origin in several EU Member States[11] (European Commission, Directorate-General for Health and Consumers 2011).

Overall, FVO's findings have revealed that the audited Member States implemented Regulation 669/2009 to a satisfactory extent, except for some minor shortcomings concerning:

- Prior notification of CED by business operators,
- Onward transportation (in particular, inefficient communication between competent authorities of different Member States),
- Respect of control frequencies for physical checks set in Annex I.

Similarly, subsequent audits performed in other Member States during the period 2012–2013 have shown a satisfactory level of implementation of the provisions of Regulation 669/2009. However, onward transportation and cooperation between competent authorities still stand out as areas requiring improvements.[12]

2.8 Annex I to Regulation (EC) No. 669/2009

Staff of DPE must conduct reinforced controls on imports listed in Annex I to Regulation 669/2009 at the control intensity therein specified. The Regulation requires regular reviews of the Annex to be carried out, at least, on a quarterly basis (Article 2, last sentence). Based on that, Annex I was updated several times since the entry into application of the Regulation. Table 2.2 provides a chronological summary of all the legislative amendments of Annex I occurred so far.

2.8.1 Scope

By definition, the list of imports contained in Annex I covers feed and food of non-animal origin that present a known or emerging risk. Groundnuts, tropical fruits, vegetables, herbs and spices are the products known to recur more often in

---

[11] The series of audits covered, in particular, the following countries: Sweden, Luxembourg, Denmark, Romania, Italy, Bulgaria, France, United Kingdom, Belgium, the Netherlands, Germany and Lithuania. A previous series of audits on import controls on food of non-animal origin was conducted over the period 2006–2008 (European Commission, Directorate-General for Health and Consumer Protection 2009).

[12] During 2012, the FVO performed audits also in Greece and Poland, in addition to a follow-up mission to Bulgaria. During 2013, the FVO visited Austria, Czech Republic and Hungary.

Table 2.2 Quarterly reviews of Annex I to Regulation 669/2009 (Sect. 2.8)

Amendment	Entry into application	Amendment	Entry into application
Reg. (EU) No. 878/2010	1 October 2010	Reg. (EU) No. 1235/2012	1 January 2013
Reg. (EU) No. 1099/2010	1 January 2011	Reg. (EU) No. 270/2013	1 April 2013
Reg. (EU) No. 187/2011	1 April 2011	Reg. (EU) No. 618/2013	1 July 2013
Reg. (EU) No. 433/2011	1 July 2011	Reg. (EU) No. 925/2013	1 October 2013
Reg. (EU) No. 799/2011	1 October 2011	Reg. (EU) No. 1355/2013	1 January 2014
Reg. (EU) No. 1277/2011	1 January 2012	Reg. (EU) No. 323/2014	1 April 2014
Reg. (EU) No. 294/2012	1 April 2012	Reg. (EU) No. 718/2014	1 July 2014
Reg. (EU) No. 514/2012	1 July 2012	Reg. (EU) No. 1021/2014	1 October 2014
Reg. (EU) No. 889/2012	1 October 2012	Reg. (EU) No. 1295/2014	1 January 2015

the Annex. For each product or group of products listed, Annex I specifies whether the relevant entry refers to products destined to animal nutrition, human consumption or both of them. The frequency required for the performance of identity and physical checks (5, 10, 20 or 50 %) by DPEs is also spelled out. To date, the hazards that are most commonly targeted in Annex I include:

- Mycotoxins (aflatoxins and Ochratoxin A)
- Pesticide residues
- Heavy metals
- Microbiological contamination (e.g. *salmonella spp*, norovirus and hepatitis A).

However, in this respect, it should be noted that the range of product/hazard combinations that may be subject to inclusion in Annex I is virtually unlimited. This means that other products of non-animal origin (e.g. food supplements, improvement agents, novel foods or feed and food containing unauthorised 'genetically modified organisms') and other hazards might as well be included in Annex I in future. In one of the most recent reviews of Annex I, in fact, enzymes of Indian origin were listed for the very first time for possible presence of veterinary residues (European Commission 2014d).

2.8.2 Listing and Quarterly Updates

Regulation 669/2009 enumerates the information sources that may be used for the inclusion of imports in Annex I, although in a not exhaustive manner (Article 2). Those sources include:

- Information reported by Member States and other associated countries (e.g. Norway, Iceland etc.) through the RASFF
- Outcome of FVO audits in non-EU countries
- Any report, information and other assurances supplied by concerned non-EU countries

- Data and information supplied by EU Member States or exchanged between them, the European Commission and the European Food Safety Authority (EFSA)
- Any relevant scientific assessment or information (e.g. EFSA opinions).

As already anticipated, results of border controls that Member States regularly submit to the European Commission constitute other key information to assess compliance levels of listed imports. The Commission makes regularly available annual reports on results of border controls to stakeholders and the general public.[13]

Current practice shows that listed imports tend to remain subject to reinforced border surveillance for at least 6 months before controls can be lifted, provided that satisfactory levels of compliance are consistently reported by Member States.

Interestingly, recital 3 of Regulation 669/2009 makes reference to a standardised methodology for the setting of Annex I. Despite some attempts, said methodology has not been established. Indeed, the difficulties in setting a rigorous methodology to apply when reviewing Annex I are apparent if one considers all the variables that may come into play for any given listing (e.g. nature of the product, risk, number of RASFF notifications and severity of the findings, outcome of FVO audits, and trade volumes). Those difficulties eventually led the Commission and the Member States to embrace a case-by-case approach when reviewing Annex I—approach that, in any event, is always based on expert assessment.

2.8.3 *From Reinforced Controls to Emergency Measures*

During 2012, the European Commission identified, among the imports listed in Annex I, a number of products for which compliance levels worsened or did not point to any significant improvement. Consequently, based on quarterly results of official controls and RASFF notifications, the Commission decided to impose additional import conditions on such products.

As a result, the following products have been included in Regulation (EU) No. 91/2013—now replaced by Regulation (EU) No. 885/2014 (European Commission 2013g, 2014f)—an emergency measure which is applicable as of 18 February 2013:

- Groundnuts and derived products from Ghana and India for possible aflatoxins contamination
- Watermelon seeds from Nigeria for possible aflatoxins contamination
- Curry leaves and okra from India for possible presence of pesticide residues.

---

[13] EU reports containing consolidated results on official controls performed by EU Member States and Norway pursuant to Regulation 669/2009 for the period 2010–2013 are available at http://ec.europa.eu/food/food/controls/increased_checks/index_en.htm.

Under this emergency measure, whereas a certain level of official controls is maintained at EU borders, business operators are required to provide, in addition to a CED, the following documents:

- A valid health certificate, signed and stamped by the competent authorities of the exporting country, and
- Analytical results of laboratory tests performed in accordance with EU requirements and prior to export in the country of origin.

Overall, the tightening of control requirements for products presenting high levels of non-compliance appears consistent with the approach that the EU applies to imports of non-animal origin, whereby the intensity of surveillance must be proportionate to the seriousness of the risk involved.

Because of the regular reviews to which Annex I is subject, very few imports have been listed since the entry into force of the Regulation: moreover, in most cases, they have undergone changes entailing, e.g. decreasing or increasing of the frequency for physical inspections. In only one case—vine fruit from Uzbekistan for possible contamination with ochratoxin A—the relevant control requirements have remained systematically unaltered. The discontinuity observed in the trade patterns concerning this product as well as the inconsistencies in compliance levels may explain why the control frequency for physical checks initially set (i.e. 50 %) has not been modified so far.

2.9 EU Reinforced Border Surveillance in the Context of Multilateral Trade Rules

By establishing a system designed to counteract public health risks that are associated with imports of feed and food of non-animal origin, Regulation 669/2009 is to be considered a 'sanitary measure' within the meaning of the Sanitary and Phytosanitary (SPS) Agreement of the World Trade Organisation (WTO).

In accordance with the provisions on transparency of the SPS Agreement (Article 7 and Annex B), the EU duly notified the draft Regulation to the SPS Secretariat and WTO members (G/SPS/N/EEC/341 of 28 April 2009).

Overall, the main criticisms raised by EU trade partners in relation to the reinforced surveillance mechanism introduced by the Regulation revolved around the lack of:

- Clear and objective conditions upon which listing of imports of non-animal origin in Annex I and subsequent modifications, including de-listing, are decided
- Maximum timelimits for the performance of physical checks (in relation to which the Regulation only requires the national competent authorities to carry them out 'as soon as technically possible').

With regard to both issues, the European Commission has provided further clarifications within the relevant Guidance document (Sect. 2.8), Q&A n. 8, 9, 10 and 25.

Interestingly, a few new listings ensuing from the quarterly reviews of Annex I led some WTO members to voice their concerns directly at official meetings of the SPS Committee.[14]

In March 2012, for instance, China has expressed strong concerns over the listing in Annex I of its exports of noodles due to the unauthorised presence of aluminium. Besides complaining about the adverse impact of border controls on trade, China openly contested the maximum limit (10 mg/kg) set by the EU for presence in food of that substance. The limit in question was based on a scientific opinion that EFSA issued in 2008 (EFSA AFC 2008) and determined in a way that the average weekly intake would be inferior to 1 mg/kg per bodyweight. China argued, instead, the weekly intake being below 2 mg/kg, following the later advice by the FAO/WHO Joint Expert Committee on Food Additives of the *Codex Alimentarius*. Notwithstanding this divergence of views on the maximum limit allowed for aluminium, according to the EU, the high non-compliance rate emerging from official controls as well as the severity of the findings reported (in some cases up to 50 mg/kg) fully justified an increased level of border surveillance.

A year earlier, concerns over the implementation of Regulation 669/2009 by EU Member States had been raised by the Dominican Republic. At that time, this country had several tropical fruits and vegetables subject to reinforced checks because of the occurrence of pesticide residues. The Dominican authorities referred that some of their products (i.e. bananas and mangoes) continued to be subject to an increased level of checks at EU borders, despite the recent lifting of the relevant control requirements.

References

Alemanno A (2010) L'approche européenne de la sécurité des importations—Concilier protection des consommateurs et accès au marché après l'affaire du lait chinois frelaté. Rev Droit Union Europ 3:527–548

Council of the European Communities (1991) Council Directive 91/496/EEC of 15 July 1991 laying down the principles governing the organization of veterinary checks on animal entering the Community from third countries and amending Directives 89/662/EEC, 90/425/EEC and 90/675/EEC. Off J Eur Comm L268:56–78

---

[14] The SPS Committee is an intergovernmental body established by the SPS Agreement (Article 14). Any WTO member has the right to be represented at meetings of the Committee. International organisations such as *Codex Alimentarius*, World Animal Health Organization and the International Plant Protection Convention enjoy the status of permanent observers within the Committee. Other international organisations, such as the Organisation for Economic Co-operation and Development (OECD) and the International Organisation for Standardisation (ISO), may take part in meetings as observers on ad-hoc basis. The SPS Committee is a forum designed to monitor and ensure the appropriate application of the SPS Agreement. In such a context, it encourages and facilitates the exchange of views between WTO members on animal health, food safety and plant health issues. It can also elaborate procedures and guidance in order to ensure WTO members fulfil the obligations stemming from the SPS Agreement. The Committee normally meets three times per year in Geneva.

Council of the European Union (1997) Council Directive 97/78/EC of 18 December 1997 laying down principles governing the organisation of veterinary checks on products entering the Community from third countries. Off J Eur Comm L24:9–30

EFSA AFC (2008) Scientific opinion of the panel on food additives, flavourings, processing aids and food contact materials on a request from European Commission on Safety of aluminium from dietary intake. EFSA J 754:1–34. Available http://www.efsa.europa.eu/en/efsajournal/doc/754.pdf. Accessed 24 Sep 2014

European Commission (2000) White paper on food safety. Brussels, 12.1.2000 COM (1999) 719 final

European Commission (2008) Commission Decision 2008/47/EC of 20 December 2007 approving the pre-export checks carried out by the United States of America on peanuts and derived products thereof as regards the presence of aflatoxins. Off J Eur Union L11:12–16

European Commission (2009a) Commission Regulation (EC) No. 669/2009 of 24 July 2009 implementing Regulation (EC) No. 882/2004 of the European Parliament and of the Council as regards the increased level of official controls on imports of certain feed and food of non-animal origin and amending Decision 2006/504/EC. Off J Eur Union L194:11–21

European Commission (2009b) Commission Regulation (EC) No. 1152/2009 of 27 November 2009 imposing special conditions governing the import of certain foodstuffs from certain third countries due to contamination risk by aflatoxins and repealing Decision 2006/504/EC. Off J Eur Union L313:40–49

European Commission, Directorate-General for Health and Consumer Protection (2009) Overview report on a series of missions in certain Member States to assess the control measures in place for import controls on food of non-animal origin and to monitor compliance with Commission decisions on the import of certain products regarding mycotoxins and Sudan Dyes adulteration'. DG (SANCO)/2009-8328. Available http://ec.europa.eu/food/fvo/specialreports/2011_6577_en.pdf. Accessed 28 Aug 2014

European Commission (2010a) Commission Regulation (EC) No. 258/2010 of 25 March 2010 imposing special conditions on the imports of guar gum originating in or consigned from India due to contamination risks by pentachlorophenol and dioxins, and repealing Decision 2008/352/EC. Off J Eur Union L80:28–31

European Commission (2010b) Commission Decision 2010/313/EU of 7 June 2010 authorising physical checks pursuant to Regulation (EC) No. 669/2009 to be carried out at approved premises of feed and food business operators in Cyprus. Off J Eur Union L140:28–29

European Commission (2010c) Commission Decision 2010/458/EU of 18 August 2010 authorising physical checks pursuant to Regulation (EC) No. 669/2009 to be carried out at approved premises of feed and food business operators in Malta. Off J Eur Union L218:26–27

European Commission (2010d) Report from the Commission to the European Parliament and the Council on the effectiveness and consistency of sanitary and phytosanitary controls on imports of food, feed, animals and plants. 21.10. 2010, COM (2010) 785 final

European Commission, Directorate-General for Health and Consumers (2011) Final Overview report of a series of audits carried out in 2010 and 2011 in certain Member States in order to assess the official control system in place for import controls on food of non-animal origin. DG (SANCO)/2011-6577. Available http://ec.europa.eu/food/fvo/specialreports/gr_2009-8328_aw_en.pdf. Accessed 28 Aug 2014

European Commission (2011a) Commission Implementing Decision 2011/402/EU of 6 July 2011 on emergency measures applicable to imports of fenugreek seeds and certain seeds and beans from Egypt. Off J Eur Union L179:10–12

European Commission (2011b) Commission Implementing Decision 2011/718/EU of 28 October 2011 amending Implementing Decision 2011/402/EU on emergency measures applicable to fenugreek seeds and certain seeds and beans imported from Egypt, Off J Eur Union L285:53–55

European Commission (2011c) Commission Regulation (EU) No. 284/2011 of 22 March 2011 laying down specific conditions and detailed procedures for the import of polyamide and melamine plastic kitchenware originating in and consigned from the People's Republic of China and Hong Kong Special Administrative Region, China. Off J Eur Union L77:25–29

European Commission (2011d) Commission Implementing Regulation (EU) No. 844/2011 of 23 August 2011 approving the pre-export checks carried out by Canada on wheat and wheat flour as regards the presence of ochratoxin A. Off J Eur Union L218:4–7

European Commission (2013a) Report from the Commission to the European Parliament and to the Council on the overall operation of official controls in the Member States on food safety, health and animal welfare and plant health, 4.10.2013, COM (2013) 681 final

European Commission (2013b) Proposal for a Regulation of the European Parliament and the Council on Animal Health, COM (2013) 260 final

European Commission (2013c) Proposal for a Regulation of the European Parliament and the Council on the production and making available on the market of plant reproductive material, COM (2013) 262 final

European Commission (2013d) Proposal for a Regulation on protective measures against pests of plants, COM (2013) 267 final

European Commission (2013e) Proposal for a Regulation of the European Parliament and the Council on official controls and other official activities performed to ensure the application of food and feed law, rules on animal health and welfare, plant health, plant reproductive material, plant protection products, COM (2013) 265 final

European Commission (2013f) Commission Regulation (EU) No. 211/2013 of 11 March 2013 on certification requirements for imports into the Union of sprouts and seeds intended for the production of sprouts. Off J Eur Union L38:26–29

European Commission (2013g) Commission Implementing Regulation (EU) No. 91/2013 of 31 January 2013 laying down specific conditions applicable to the import of groundnuts from Ghana and India, okra and curry leaves from India and watermelon seeds from Nigeria and amending Regulations (EC) 669/2009 and (EC) 1152/2009. Off J Eur Union L33:2–10

European Commission (2014a) Commission Implementing Decision 2014/88/EU of 13 February 2014 suspending temporarily imports from Bangladesh of foodstuffs containing or consisting of betel leaves ('Piper betle'). Off J Eur Union L45:34–35

European Commission (2014b) Commission Implementing Decision 510/2013/EU of 29 July 2014 amending Implementing Decision 2014/88/EU suspending temporarily imports from Bangladesh of foodstuffs containing or consisting of betel leaves ('Piper betle') as regards its period of application. Off J Eur Union L 228:33–34

European Commission (2014c) Commission Regulation (EU) No. 718/2014 of 27 June 2014 amending Regulation (EC) No. 669/2009 implementing Regulation (EC) 882/2004 as regards an increased level of official controls on import of certain feed and food of non-animal origin. Off J Eur Union L190:55–61

European Commission (2014d) Commission Implementing Regulation (EU) No. 323/2014 of 28 March 2014 amending Annexes I and II to Regulation (EC) No. 669/2009 implementing Regulation (EC) No. 882/2004 of the European Parliament and of the Council as regards the increased level of official controls on imports of certain feed and food of non-animal origin. Off J Eur Union L35:12–23

European Commission (2014e) Commission Implementing Regulation (EU) No. 884/2014 of 13 August 2014 imposing special conditions governing the import of certain feed and food from certain third countries due to contamination risk by aflatoxins and repealing Regulation (EC) No. 1152/2009. Off J Eur Union L 242:4–19

European Commission (2014f) Commission Implementing Regulation (EU) No. 885/2014 of 13 August 2014 laying down specific conditions applicable to the import of okra and curry leaves from India and repealing Implementing Regulation (EU) No. 91/2013. Off J Eur Union L 242:20–26

European Parliament and Council (2002) Regulation (EC) No. 178/2002 of the European Parliament and of the Council of 28 January 2002 laying down the general principles and requirements of food law, establishing the European Food Safety Authority and laying down procedures in matters of food safety. Off J Eur Comm L31:1–24

European Parliament and Council (2004) Regulation (EC) No. 882/2004 of the European Parliament and of the Council of 29 April 2004 on official controls performed to ensure the verification of compliance with feed and food law, animal health and welfare rules. Off J Eur Union L165:1–141

González Vaqué L (2013a) Los Estados miembros de la UE como gestores de los controles oficiales relativos a la seguridad de los alimentos, la salud y el bienestar de los animales y la fitosanidad. Riv Dirit Aliment 4:20–28. Available http://www.rivistadirittoalimentare.it/rivista/2013-04/VAQUE.pdf. Accessed 27 Aug 2014

González Vaqué L (2013b) La Propuesta de la Comisión Europea para simplificar, racionalizar y uniformizar los controles sobre los alimentos: ¿cómo afectará al conjunto de la la legislación alimentaria? Riv Dirit Aliment 2:14–22. Available http://www.rivistadirittoalimentare.it/rivista/2013-02/VAQUE.pdf. Accessed 27 Aug 2014

Laurenza E (2013) The EU's new regulatory framework on official controls, animal health, plant health and seeds. Eur J Risk Regul 3:381–383

Montanari F (2013) Controlli ufficiali - La proposta di riforma della Commissione UE. Alimenti Bevande 6:22–25

Paganizza V (2013) Les quatres mosquetaires (ou mosquetons) contre *E. Coli*: I Regolamenti (UE) 208/2013. 209/2013, 210/2013, 211/2013 e gli eccessi nella sicurezza. Riv Dirit Aliment 2:36–44. Available http://www.rivistadirittoalimentare.it/rivista/2013-02/PAGANIZZA.pdf. Accessed 27 Aug 2014

Rodriguez Font M (2012) The 'Cucumber Crisis': legal Gaps and lack of precision in the risk analysis system in food safety. Riv Dirit Aliment 2:1–15. Available http://www.rivistadirittoal imentare.it/rivista/2012-02/RODRIGUEZ%20FONT.pdf. Accessed 27 Aug 2014

Chapter 3
Managing Risks in Imports of Non-animal Origin: EU Emergency Measures

Veronika Jezsó

Abstract This chapter describes the current set of emergency measures, also known as 'safeguard measures', on which the European Union (EU) relies upon for addressing high risks in imports of feed and food of non-animal origin. Overall, EU emergency measures involve the applicability of specific import conditions and control activities of certain products of non-animal origin from certain non-EU countries that are destined to the internal market. Following a brief introduction of emergency measures and the legal framework in the EU, this chapter provides a detailed overview of the measures currently in force in the area of imports of non-animal origin. The discussion concerns long-standing measures, such as those addressing aflatoxin contamination in imports of dried fruits, and more recent ones, such as the temporary ban that the EU has introduced on betel leaves from Bangladesh. Other relevant measures are also discussed in this chapter with relation to the area of imports of non-animal origin, including measures following the *Fukushima* nuclear accident or imports of GMO rice products from China.

Keywords Emergency measure · Feed · Food · Import control · Import conditions · Non-animal origin

Abbreviations

ARfD	Acute Reference Dose
BIP	Border Inspection Post
Cs-134	Caesium-134
Cs-137	Caesium-137
CN	Combined Nomenclature
CED	Common Entry Document
CP	Control Point
CVED	Common Veterinary Entry Document
DPE	Designated Point of Entry

© The Author(s) 2015
F. Montanari et al., *Risk Regulation in Non-Animal Food Imports*,
Chemistry of Foods, DOI 10.1007/978-3-319-14014-8_3

DPI	Designated Point of Import
EFSA	European Food Safety Authority
EU	European Union
FVO	Food and Veterinary Office
DG SANCO	Directorate-General for Health and Consumers
GFL	General Food Law
GMO	Genetically modified organism
MRL	Maximum Residue Level
PCP	Pentachlorophenol
RASFF	Rapid Alert System for Food and Feed
SCoFCAH	Standing Committee on Food Chain and Animal Health
UK	United Kingdom

3.1 Emergency Measures: An Introduction

When confronted with a serious risk, the European Union (EU) may rely upon different policy tools, including crisis management procedures and legislative measures. With relation to the EU, these emergency actions—also known as 'safeguard measures'—correspond to one of the most effective instruments for the prevention, reduction or eradication of risks for public, animal and plant health (Sect. 2.3.3).

Since risks associated with emergency situations can be of different nature, the EU must be, each time, in the position to tailor an appropriate response in terms of risk management. Therefore, when it comes to imports, requirements laid down in EU emergency measures may be diverse and include controls performed prior to export, provision of official certificates, prior notification of the arrival of the consignment as well as controls at import.

Generally, EU import requirements set in emergency measures are established based on the risk categorisation in relation to the concerned products (Sect. 2.3.3). The following factors should be, *inter alia*, taken into account:

- Information on food incidents, contaminations or outbreaks of diseases and pests
- Data on trade exchanges
- Notifications or interceptions of unsafe or otherwise non-compliant products
- Scientific assessment and correlated evidence
- Guarantees provided by non-EU countries, including applicable legislation, safety standards and the existing system of official controls.

In accordance with EU legislations on imported goods, risks involved are generally addressed by developing a response that is proportionate to the seriousness of the possible threat to human, animal or plant health. In other words, the higher the risk presented by a given import, the stricter the conditions for its entry into the EU are likely to be.

Once an emergency measure is in place, the necessary monitoring and verification are ensured at EU level. As a result, the evolution of the risk is closely followed. Due to the diversity of circumstances, reasons and risks that may underpin emergency measures, different scenarios may be observed and the evolution of the single situation cannot be easily predicted, in principle. It may be concluded that a particular emergency measure is finally lifted when the ensuing health threat has thus ceased to exist.

The European Commission holds regular discussions with Member States, within relevant expert groups and regulatory committees, on emerging or reoccurring risks. The aim of these meetings is to develop, if need be, appropriate risk management responses, including the adoption of emergency measures. However, the Commission may take immediate and independent measures without a preliminary discussion, pending confirmation by EU Member States, if circumstances require an urgent response.

3.2 EU Emergency Measures on Imports of Non-animal Origin: The Relevant Legislative Framework

Regulation (EC) No. 178/2002—commonly known as the 'General Food Law' (GFL)—lays down general principles governing EU food policy (European Parliament and Council 2002). In general, the free movement of feed and food products within the EU market has to be assured while protecting public health and consumer interests. In addition, feed and food imported into the EU must comply with EU safety requirements or with other equivalent standards (Sect. 2.2.1).

3.2.1 The General Food Law: Articles 53 and 54

Past incidents with imported or domestically produced food and feed have confirmed the need to have in place adequate instruments for the prevention and the management of safety risks. Therefore, Article 53 of the GFL confers special powers to the European Commission for taking emergency measures when EU or non-EU food or feed may 'likely to constitute a serious risk to human health, animal health or the environment' and Member States may not be able to contain that risk satisfactorily.

The procedure for the adoption of emergency measures can be initiated by the European Commission or upon request of a Member State. Depending on the seriousness of the risk involved, emergency measures can manifest in (Sect. 2.2.3):

- Suspension of the imports in question from all or part of the exporting non-EU country and, where relevant, from the non-EU country where such goods transit through
- Special conditions for import from all or part of the concerned non-EU country
- Any other appropriate interim measure.

Should circumstances require urgent actions, the European Commission may provisionally adopt emergency measures after consulting with the Member State(s) concerned and informing the other Member States. Should this be the situation, adopted decisions need to be confirmed, amended, revoked or extended by the relevant section of the Standing Committee of the Food Chain and Animal Health (SCoFCAH) within 10 days.

In addition, the GFL also foresees in certain cases the possibility for Member States to adopt interim protective measures at national level (Article 54) on condition that adequate information is sent to the Commission and other Member States.

National measures are then subject to scrutiny of the SCoFCAH, which can rule on their extension, modification or repeal. The Member State adopting temporary protective measures may maintain these provisions until an EU decision is formally taken. It has to be noted, however, that EU legislative practice reveals to date a limited use of Article 54 as opposed to Article 53 of the GFL.

3.2.2 The GFL and Regulation (EC) No. 882/2004

The GFL is complemented by Regulation (EC) No. 882/2004 (Sect. 2.2.2, European Parliament and Council 2004). With reference to imported food and feed, paragraphs 1 and 2 of Regulation 882/2004, Article 48, contain a provision that may recall the wording of the GFL, Article 53, to some extent. In detail, specific import conditions on products originating from non-EU countries may be determined, including (Sect. 2.2.2):

(a) The establishment of lists of non-EU countries allowed to export specific products to the EU
(b) The use of specific certificates for consignments
(c) Special import conditions depending on the products and correlated risks.

Although partially overlapping, the main difference between this provision and Article 53 of the GFL appears to rest on the different rationale. In other terms, emergency measures under the GFL, Article 53, are, in principle, limited in time, while import conditions under Reg. 882/2004, Article 48, are meant to be permanent.

As regards imports of non-animal origin, use of Article 48 of Reg. 882/2004 has been overall relatively limited at EU level. Conversely, Article 53 of the GFL is currently the legal basis most frequently used at EU level in this area, although its choice may be questionable in certain instances.

3.3 EU Emergency Measures on Imports of Non-animal Origin: General Structure

Generally, EU emergency measures concerning imports of non-animal origin consist of a text rather short and concise and a few annexes. They usually take the form of a *regulation* or a *decision*. Since under EU law *regulations* are legal acts of general application, emergency measures take that form whenever rights and obligations for different interested parties (e.g. Member States' competent authorities, feed and food business operators) are foreseen. On the other hand, *decisions* produce their effects only towards specific addressees. They are therefore mostly employed for the adoption of emergency measures imposing, e.g. prohibition of imports, which, as such, are merely directed at (and thus to be implemented by) Member States' competent authorities.

Because of the specific nature of imports and their origins, the scope of emergency measures must necessarily be well defined. Targeted product groups are usually associated with codes of the customs Combined Nomenclature[1] (CN) in order to allow their precise identification.

With concern to the entry into force of emergency actions, the date is explicitly spelt out in the text, as a general rule. In some other cases, however, measures apply 20 days following their publication in the EU Official Journal.

3.3.1 Import Conditions

Emergency measures normally lay down a set of specific import requirements, including possible import conditions and/or controls (Sect. 2.3.5). In detail, the following documents or certificates (produced prior to dispatch) are usually requested when a particular food or feed, falling under the scope of EU import conditions, arrive at the EU borders:

- An analytical report, indicating the favourable results of official sampling and analysis carried out on the products prior to their export and issued by an accredited laboratory in the exporting country, in accordance with the applicable EU requirements in the area
- A health certificate issued by an authorised representative of the competent authority in the exporting non-EU country. This document certifies that the product intended to be imported into the EU is in compliance with the applicable EU requirements. Normally, the validity period of this document is 4 months
- A declaration issued by an authorised representative of the competent authority in the exporting non-EU country. This document attests specific information required to the effect of the emergency measure, e.g. that the product complies with the legislation in force in the exporting non-EU country.

---

[1] When declared to customs in the EU, goods must generally be classified according to the combined nomenclature or CN. Imported and exported goods have to be declared stating under which subheading of the nomenclature they fall. This determines which rate of customs duty applies and how the goods are treated for statistical purposes. Useful information may be available at the following web address: http://ec.europa.eu/taxation_customs/customs/customs_duties/tariff_aspects/combined_nomenclature/index_en.htm.

3.3.2 Import Controls

On the other hand, import controls involve surveillance activities or checks to be performed by competent authorities in EU Member States before products can be eventually imported.

Feed and food business operators are usually required to provide prior notification of the estimated date and time of the arrival of their goods. The prior notification must be given to the point of entry in the EU that is expressly set out for such a purpose by each emergency measure, e.g. the Designated Point of Entry (DPE). In this respect, the Common Entry Document (CED), as established by Annex II of the Regulation (EC) No. 669/2009 (European Commission 2009a), is increasingly used in the area of imports of non-animal origin for the purpose of prior notification.

Official controls involve generally documentary, identity and physical checks. Typically, documentary checks are systematic (i.e. 100 % on all relevant consignments), while identity and physical checks are to be conducted based on the frequency set in the measure.

Whenever, following such controls, cases of non-compliance are detected, Member States' competent authorities must undertake actions in accordance with relevant provision of Regulation 882/2004 (Articles 19, 20 and 21). In addition, Member States' competent authorities must inform the European Commission via the Rapid Alert System for Food and Feed (RASFF) in accordance with the GFL, Article 50, when detected non-compliances involve serious risks or a rejection at a EU border.

In all cases, consignments must remain under official supervision until the feed and food business operator decides to offer them for import and present them to customs for release for free circulation. In order to do so, the operator must provide customs authorities with documented evidence (e.g. CED) that the sanitary official controls have been carried out and, where applicable, that the results of physical checks have been favourable.

3.3.3 Splitting of Consignments

In some cases, feed and food business operators might want their own consignments to be split in order to forward relevant parts or fractions to different destinations within the EU. This situation may occur when large consignments are delivered. When speaking of emergency measures, this procedure is not generally allowed until the end of all required checks. The subsequent splitting of a consignment is, however, allowed as long as a certified copy of the document—attesting the results of official controls—accompanies each part or fraction until its release for free circulation.

3.3.4 Costs

In all emergency measures regarding imports of non-animal origin into the EU, costs resulting from the performance of official controls (including sampling, analysis, storage and any measures taken following non-compliance) must be borne by the feed and food business operator.

3.3.5 Quarterly Reporting

Several emergency measures require EU Member States to submit a report on a regular basis, detailing the results of control activities to the European Commission. The report is usually expected to be provided every 3 months (Sect. 2.5) and must give account of the following:

- The number of received consignments
- The number of consignments subjected to sampling for analysis
- The results of checks.

3.4 EU Emergency Measures: Causes

The adoption of EU emergency measures on imports of non-animal origin is usually triggered by one or more factors, most commonly (Fig. 3.1) the following:

- Food incidents and contaminations
- Anthropic accidents (e.g. *Fukushima* nuclear power station) or natural disasters

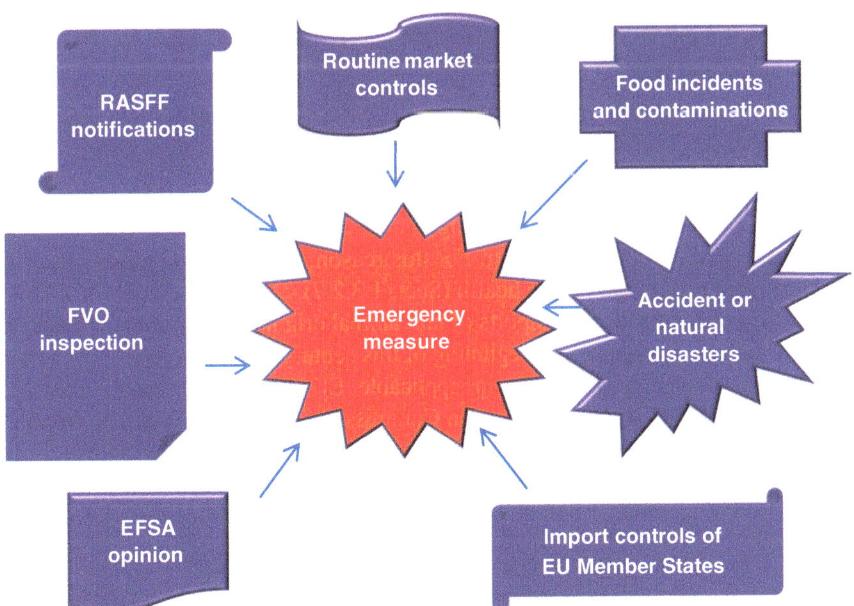

Fig. 3.1 Main elements that can trigger emergency measures related to imports of non-animal origin

- Number of RASFF notifications and levels of non-compliance
- Results from 'routine' official control activities performed on imports on the market
- Findings of audits conducted in non-EU countries by the Food and Veterinary Office (FVO) of the European Commission (Sect. 2.7)
- Any information received from EU Member States, concerned non-EU countries or stakeholders
- Scientific opinions, reports or any other scientific evidence of the European Food Safety Authority (EFSA) or other relevant scientific organisations (Sect. 2.8.2).

Above-mentioned elements can justify the adoption of emergency measures either alone or in combination with other concurring factors. Most commonly, the confirmation on the nature and the extent of the emergency situation is sought through targeted investigations, as soon as circumstances point out to the possible presence of a health risk. For example, the performance of dedicated 'ad hoc' FVO audits may be foreseen in order to gather relevant information on the official control system in place in the country exporting to the EU and/or to determine the origin of a contamination.

3.5 Emergency Measures on Imports of Non-animal Origin

3.5.1 Aflatoxin Contamination in Nuts and Dried Fruits from Various Non-EU Countries: Regulation (EU) No. 884/2014

3.5.1.1 Background

Aflatoxin is a potent genotoxic carcinogen and, even at extremely low levels, can contribute to the risk of liver cancer. For this reason, aflatoxin contamination constitutes a serious threat to public health (Sect. 1.3.2.7).

Aflatoxin contamination in imports of non-animal origin is a long-standing issue at EU level. Indeed, as early as the beginning of this century, high exceedances of maximum levels of aflatoxins set out in applicable EU legislation at that time (i.e. Regulation (EC) No. 466/2001, European Commission 2001) were regularly observed in relation to certain food products (mainly nuts, pistachios and figs) originating from various non-EU countries. This led EU decision-makers to adopt a set of different—though similar in terms of purpose and content—emergency measures for each product(s)/non-EU country combination (European Commission 2000, 2002a, b, 2003, 2005) and subsequent amendments[2] (European Commission 2004a, b).

With Regulation (EC) No. 882/2004 and the consequent harmonisation in the organisation of official controls, all emergency measures concerning aflatoxins in imports of non-animal origin have been merged under a single legislative framework, Decision 2006/504/EC (European Commission 2006a). This legislation covered the following products from various non-EU countries:

- Brazil: Brazil nuts in shell, mixtures of nuts or dried fruits containing Brazil nuts
- China: (roasted) peanuts
- Egypt: (roasted) peanuts
- Iran: (roasted) pistachios
- Turkey: dried figs, hazelnuts, pistachios, mixtures of nuts, fig paste and hazelnut paste, etc.

Overall, official controls performed by EU Member States under Decision 2006/504/EC have repeatedly revealed non-compliances with maximum levels in accordance with Regulation (EC) No. 1881/2006 (European Commission 2006b). In addition, FVO audits in concerned non-EU countries have signalled several deficiencies in their national official control systems.

For these reasons, Regulation (EC) No. 1152/2009 had eventually replaced Decision 2006/504/EC (European Commission 2009b). In particular, Regulation No. 1152/2009 had introduced, among others, a few new products (e.g. almonds and mixtures from the United States of America) and reviewed control intensities for physical inspections for certain other commodities. Most importantly, however, it introduced a new system for official controls consisting of DPE, as amended by Regulation (EU) No. 91/2013 (European Commission 2013a) and Designated Points of Import (DPI).

Subsequently, the EU regime on aflatoxin risk in imports of non-animal origin has been thoroughly reviewed again, following a minor amendment by Regulation

---

[2] In detail, emergency measures and subsequent amendments can be chronologically summarised as follows:

- With reference to the import of peanuts and certain products derived from peanuts originating in or consigned from Egypt, Commission Decision 2000/49/EC has been lastly amended by the Decision 2004/429/EC, corrected by in the Official Journal L189, p. 13
- With concern to the import of peanuts and certain products derived from peanuts originating in or consigned from China, Commission Decision 2002/79/EC has been amended by the Decision 2004/429/EC
- With relation to the import of figs, hazelnuts and pistachios and certain products derived thereof originating in or consigned from Turkey, Commission Decision 2002/80/EC has been amended by the Decision 2004/429/EC
- With relation to the import of Brazil nuts in shell originating in or consigned from Brazil, Commission Decision 2003/493/EC has been amended by the Decision 2004/428/EC
- With concern to import of pistachios and certain products derived from pistachios originating in or consigned from Iran, Commission Decision 2005/85/EC applied.

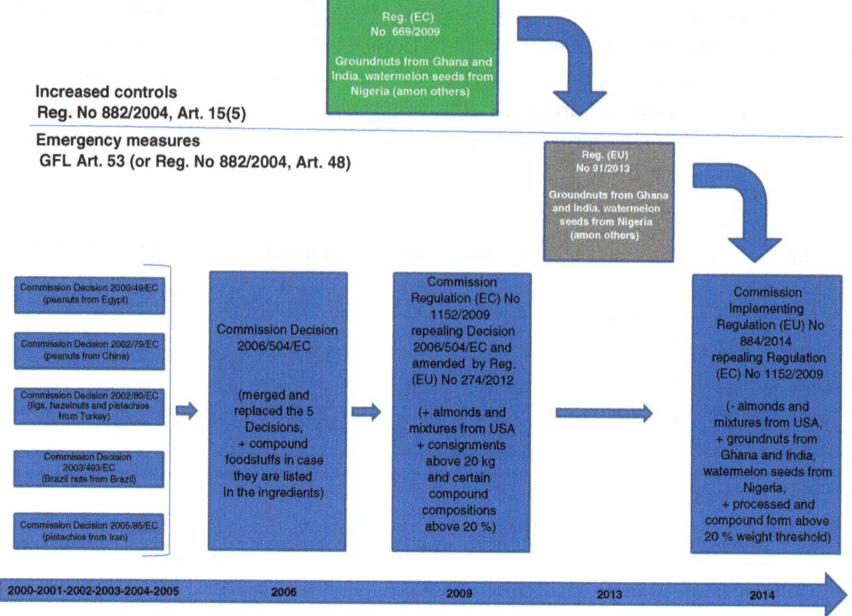

Fig. 3.2 Chronology of EU emergency measures on imports of non-animal origin presenting high risk of aflatoxin contamination

(EU) No. 274/2012 (European Commission 2012a). Indeed, the new Regulation (EU) No. 884/2014 (European Commission 2014a) has extended the original scope of the regime to animal feed. Furthermore, results of official controls by Member States' competent authorities and of FVO audits triggered certain further changes as regards control frequencies and products covered (e.g. delisting of almonds from the United States of America). Finally, some of the products listed under Regulation 91/2013 for risk of aflatoxin contamination, namely groundnuts from India and Ghana and watermelon seeds from Nigeria, were incorporated into the new measure. Through this change, all imports of non-animal origin presenting high risk of aflatoxins are now contained in one single legislative instrument. The Regulation is applicable as of 3 September 2014. Figure 3.2 illustrates the chronology of EU emergency measures on aflatoxins since 2000.

3.5.1.2 Requirements

Regulation 884/2014 applies to the following imports, when destined for commercial purposes:

- Brazil: Brazil nuts in shell and mixtures of nuts or dried fruits containing Brazil nuts in shell (food)
- China: groundnuts and peanut butter (feed and food)

- Egypt: groundnuts and peanut butter (feed and food)
- Iran: pistachios (food)
- Turkey: dried figs, hazelnuts (*Corylus* spp.), pistachios, mixtures of nuts or dried fruits containing figs, hazelnuts or pistachios; fig paste, pistachio paste and hazelnut paste; hazelnuts, figs and pistachios, including mixtures; flour, meal and powder of hazelnuts and pistachios; and cut, sliced and broken hazelnuts
- Ghana: groundnuts and peanut butter (feed and food)
- India: groundnuts and peanut butter (feed and food)
- Nigeria: watermelon seeds and derived products (food).

In addition, import requirements apply to processed and compound imports of non-animal origin, containing one of the above-mentioned products at least in a quantity exceeding 20 %. On the other hand, imports of non-animal origin that are destined to a private person for his/her personal consumption or use only are not covered by the measure (Article 1).

3.5.1.3 Key Elements of the Measure

Relevant Definitions and Minimum Requirements for Operating DPI

Regulation 884/2014 provides some definitions that are essential for determining the place where official controls must be carried out (Article 2). Table 3.1 illustrates relevant definitions.

In practice, DPEs are typically located at EU borders (i.e. sea ports, airports or land borders), where imports first reach physically the EU territory. DPIs may also be located at EU borders, but also inland within the territory of an EU Member State (Table 3.1).

In relation to DPI, Regulation 884/2014 lays down certain minimum requirements that Member States' competent authorities must ensure, such as (Article 8, paragraph 1) the following:

- Sufficient and suitably qualified staff to perform controls
- A sheltered place for unloading and sampling

Table 3.1 Definitions for Designated Points of Import and Designated Points of Entry, in accordance with Regulation No. 884/2014, Article 2 (European Commission 2014a)

Designated Points of Import (DPI)	'means any point designated by the competent authority, through which the food or feed [...] may be imported into the [European]Union'
Designated Points of Entry (DPE)	'means the point of entry as defined in Article 3(b) of Commission Regulation (EC) 669/2009' i.e. 'means the point of entry provided for in the first indent of Article 17(1) of Regulation (EC) No. 882/2004, into one of the territories referred to in Annex I thereto; in cases of consignments arriving by sea, which are unloaded for the purposes of being loaded on another vessel for onwards transportation to a port in another Member State, the designated point of entry shall be the latter port'

- The availability of storage rooms and warehouses
- An official laboratory located at a short distance.

On the other hand, business operators must ensure that the unloading of the consignment is performed in a way that can allow representative sampling. In addition, business operators must provide the official inspector with appropriate sampling equipments in case sampling procedures are hindered because of special transport or specific packaging forms.

Finally, a full and up-to-date list of links to national DPIs has to be made available to the Commission, that is publishing this data on its website for information purposes (Article 8).

Obligations for Feed and Food Business Operators

Consignments of relevant products must be accompanied by the results of sampling and analysis performed by the competent authorities of the exporting country in order to prove compliance with EU legislation on maximum levels of aflatoxins (Article 4, paragraph 1). The sampling and analysis have to be performed in accordance with the relevant EU legislation, namely with

(a) Regulation (EC) No. 401/2006 for aflatoxins in food (European Commission 2006c), lastly amended in 2014 (European Commission 2014b), and
(b) Regulation (EC) No. 152/2009 for aflatoxins in feed (European Commission 2009c).

In addition, a valid health certificate must be provided. The certificate, signed and verified by an authorised representative of the competent authority of the exporting country, has a 4-month validity from the date of issuance (Article 5). Also, each consignment must bear an identification code on all accompanying documents as well as on each individual bag of the consignment (Article 6).

Business operators must give prenotification to the DPE about the estimated date and time of the physical arrival of the consignment. Notification must be done, at least, one working day prior to arrival, by filling the relevant part of the CED (Part I) as provided for by Regulation 669/2009 (Article 7).

Whenever the DPI is different from the DPE (i.e. located inland), the business operator is subject to an additional obligation of prior notification. This is a new requirement that Regulation 884/2014 introduced. Following the favourable outcome of documentary checks at the DPE, it may happen that, based on the business operator's decision, the consignment remains for a certain time in warehouses located within the DPE area. Under such circumstances and in order to ensure that the DPI is fully prepared when the consignment is presented for import, the business operator must notify the competent authorities of the DPI about its arrival at least one working day in advance. The notification implies sending a copy of the CED completed by the competent authority at the DPE and attesting results of the documentary checks.

Performance of Official Controls and Release for Free Circulation

Under the EU regime of reinforced border surveillance established by Regulation 669/2009, all checks (i.e. documentary, identity and physical) are conducted at DPE located at EU borders. Conversely, Regulation 884/2014 designs a different organisation for official controls performed under its framework.

As a rule, official controls must be performed within 15 working days from the moment the consignment is offered for import and physically available for sampling (Article 9, paragraph 1).

Whenever the business operator fails documentary checks, the consignment cannot enter the EU and must, instead, be redispatched to the country of origin or destroyed (Article 9, paragraph 3). If the documentary checks are favourable, the competent authorities at DPE may authorise the transfer of the consignment to the relevant DPI. The originals of the health certificate, of the results of sampling and analysis and of the CED must accompany the consignment during its transfer. In this case, the competent authorities of the DPE must inform, without delay, the competent authority of the DPI of the arrival of the consignment (Article 9, paragraph 4).

The competent authority must perform identity and physical checks at the DPI, including sampling for analysis, at a frequency set out in Annex I to Regulation 884/2014. Sampling of food must follow the provisions laid down in Annex I of Regulation 401/2006, whereas in case of feed, Annex I of Regulation 152/2009 applies.

Once the checks are completed, the competent authority at the DPI must complete, as appropriate, the relevant part of the CED (Part II), join the results of sampling and analysis when these have been performed and hand over the original of the CED, duly signed and stamped, to the business operator. Whenever the outcome of the official controls is favourable, the business operator may present this documentation to the competent customs office for obtaining the release for free circulation.

Table 3.2 provides an overview of place and type of controls in relation to imports of non-animal origin presenting a high risk of aflatoxin contamination.

Table 3.2 Place and type of controls pursuant to Regulation No. 884/2014

Place of controls	Geographical locations	Type of controls
DPE (point of first physical arrival)	EU border (i.e. seaport, airport, land border)	Documentary checks (compulsory) Identity and physical checks (possible provided that DPE is also a DPI)
DPI (any point where the consignment can be offered for import)	EU border (i.e. seaport, airport, land border) or inland	Identity and physical checks

3.5.1.4 Implementation: State of Play

EU Member States

Over the period 2010–2011, the FVO carried out a number of audits in the Member States (Sect. 2.7). The overall aim was to evaluate the implementation of the official control systems in place for imports of non-animal origin in accordance with EU law. Therefore, these audits covered implementation at national level of official controls required under EU emergency measures as well as Regulation 669/2009. The series of audits revealed that all Member States visited had adequately designated competent authorities and internal coordination within and between the sanitary competent authorities seemed to work effectively. Some deficiencies were nevertheless observed mainly in relation to the cooperation and communication between competent authorities and customs.

Finally, it should be noted that the European Commission has developed a guidance document for Member States' competent authorities in 2010, in order to encourage uniform implementation of the EU legislation related to aflatoxins. In detail, this guidance explains the provisions of the relevant EU legislation and provides practical advice in relation to different situations that may occur during enforcement.[3]

Non-EU Countries

Between 2001 and 2010, the FVO performed up to 20 audits in order to assess the official control systems in place for mycotoxin contamination in non-EU countries concerned with relevant EU measures (DG SANCO 2010). Over this period, in fact, the RASFF received more than 5,200 notifications relating to aflatoxins in nuts and dried fruits from the major exporting countries (i.e. Argentina, Brazil, China, Egypt, Ghana, India, Iran, Turkey and United States).

The audits have been carried out using document reviews and interviews with business operators and national officials, including, among others, on-site visits to business operators' premises, processing companies and shelling facilities, where official controls led by competent authorities were evaluated. Main FVO recommendations have been primarily related to sampling, accreditation of laboratories, inadequate follow-up of RASFF notifications, effectiveness of sorting techniques with regard to aflatoxin controls, traceability, participation in proficiency testing schemes and method validation.

---

[3] The current version of 'Guidance document for Competent Authorities for the control of compliance with EU legislation on aflatoxins' (last edition, November 2010) can be found at the website: http://ec.europa.eu/food/food/chemicalsafety/contaminants/guidance-2010.pdf.

A recent FVO inspection to India in 2013 (DG SANCO 2013a) (Sect. 3.5.4.2) has shown certain improvements since the previous audit in 2009, especially with reference to control arrangements for the export of peanuts and the performance of private laboratories. Recent audits carried out in Brazil and Turkey have also indicated positive trends as regards efforts to ensure compliance with EU requirements (DG SANCO 2011d, 2012). On the other hand, the report of the latest FVO inspection on China (2011) has revealed no major improvements in the control system for the prevention of aflatoxin contamination in peanuts intended for export to the EU since the previous audit (DG SANCO 2011e).

3.5.2   Pesticide Residues in Okra and Curry Leaves from India: Regulation (EU) No. 885/2014

3.5.2.1   Background

Originally, okra and curry leaves from India had been subject to the regime of reinforced border controls under Regulation 669/2009 for the presence of pesticide residues. Over two years of stepped-up border surveillance showed consistent non-compliance rates with Maximum Residue Levels ('MRLs') of plant protection products established in EU legislation and cases of serious exceedances of the Acute Reference Dose (ARfD).

Moreover, an FVO audit conducted in India (DG SANCO 2011a) revealed that, in fact, no framework was in place for the control of the use of plant protection products and presence of their residues for okra, curry leaves and other products. Following the audit, the Indian authorities failed to provide the European Commission with a satisfactory action plan to address the shortcomings identified.

Considering the situation and the risks involved for public health, the European Commission eventually adopted Regulation 91/2013, setting stricter conditions for the import of these products into the EU as well as other imports presenting high risk of aflatoxin contamination (Sect. 3.5.1.1).

However, nearly one year after the adoption of Regulation 91/2013, in the interest of an efficient and consistent organisation of official controls, the European Commission deemed appropriate to place imports presenting risk of aflatoxins and those listed for occurrence of pesticide residues under two distinct separate import regimes. In relation to pesticide residues, the Commission therefore adopted new emergency measures under Regulation (EU) No. 885/2014 (European Commission 2014c). Applicable as of 17 August 2014, this Regulation foresees the applicability of control procedures on Indian okra and curry leaves that mirror those established by Regulation 669/2009.

Regulation 885/2014 applies to fresh and frozen okra and to curry leaves, intended for human consumption, with Indian origin. It also applies to compound food that contains the above-mentioned products in a quantity above 20 %

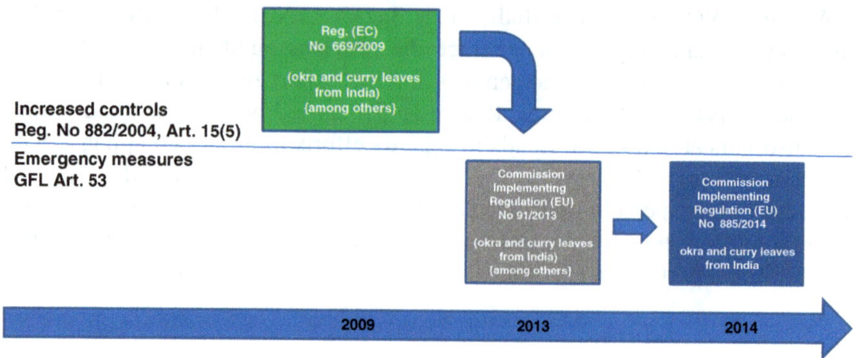

Fig. 3.3 Chronology and evolution of EU legislation on okra and curry leaves from India

(Article 1). Figure 3.3 illustrates the chronology and evolution of EU emergency measures concerning imports of okra and curry leaves from India.

3.5.2.2 Key Elements of the Measure

Obligations of Food Business Operators

Consignments can only enter the EU territory through a DPE (Article 3). Consignments must be accompanied by the results of sampling and analysis performed by the competent authorities of the exporting country in order to prove compliance with EU legislation on pesticide residues. The sampling must be performed in accordance with Directive 2002/63/EC (Article 4). In addition, a health certificate must be provided—as per the model in Annex II to the Regulation— duly signed and verified by an authorised representative of the Indian competent authorities. The certificate has four-month validity from the date of issuance (Article 5).

Concerned business operators must give a prior notification of the estimated date and time of physical arrival of consignments to the competent authorities at the DPE, by filling out the relevant part of the CED (Part I). They should equally specify the nature of the product (i.e. compound). The CED must be notified, at least, one working day prior to the arrival of the consignment (Article 7).

Performance of Official Controls and Release for Free Circulation

In terms of control activities, the competent authority at the DPE is required to carry out documentary checks on each consignment of okra and curry leaves presented for EU import. As for the identity and physical checks, the frequency of import controls is set out in Annex I to the Regulation. This latter explicitly refers to Articles 8, 9 and 19 of Regulation No. 669/2009, thereby making onward transportation, special circumstances and transitional period also applicable in the context of this emergency measure.

Following the completion of all checks, the competent authorities of the DPE must complete the relevant entries of the CED (Part II), join the results of sampling and analysis carried out when these have been performed and hand over the original of the CED duly signed and stamped to the business operator. With such documents, the business operator is entitled to present its products to the customs for obtaining the release for free circulation. Whenever the competent authorities of the DPE agree to onward transportation of the consignment, while results of physical checks are still pending, a certified copy of the original CED must be issued for that purpose (Article 8).

3.5.3 Dioxin and PCP Contamination of Guar Gum from India: Regulation (EU) No. 258/2010

3.5.3.1 Background

A serious contamination of guar gum of Indian origin with dioxins and pentachlorophenol (PCP) has been notified by the Swiss competent authorities in 2007 through the RASFF. Levels of dioxins and PCP detected in certain batches were indeed very high (i.e. about one thousand times more than the level regarded as a normal background contamination).

Guar gum is a broadly used food additive in industrial processing. In particular, it can be used as a stabiliser and emulsifier. This food additive is extracted from plants that are native of India and Pakistan. On the other hand, dioxins and PCP are organic chemical compounds with a notable series of toxic and biochemical effects; some of these molecules have been classified as known human carcinogen.

Following the Swiss RASFF notification, the FVO shortly carried out an audit in India (DG SANCO 2007). Although the audit could not identify the cause at the origin of the contamination, it revealed that the local system of official controls in place in the country was not effective enough to prevent this contamination from occurring again.

As a consequence, the European Commission introduced emergency measures by adopting Decision 2008/352/EC (European Commission 2008a). This decision required all consignments of guar gum or products containing it in significant amounts, originating or consigned from India, and destined to the EU market to be accompanied by an analytical report endorsed by the competent authority from the exporting country.

In October 2009, the FVO performed a follow-up inspection in India (DG SANCO 2009), where several deficiencies were observed and it was concluded that the contamination of guar gum with PCP and/or dioxins that occurred in 2007 could not be merely regarded as an isolated incident. As a consequence, the European Commission decided to reinforce the requirements in place for imports of guar gum and designed for this purpose a new legal framework by adopting Regulation (EU) No. 258/2010 (European Commission 2010), applicable as of 15 April 2010.

After the second audit and the consequent new strategy, the FVO has carried out a subsequent inspection in 2011, with a general level of satisfaction as regards the effectiveness of the national official control system (DG SANCO 2011b). In addition, the number of RASSF notifications concerning cases of contamination of guar gum with dioxins and PCP cannot be defined as 'high' (only 7 over the period 2007–2014). On the other hand, the repeated detection of similar contaminants reveals that the issue still raises concerns and justifies a certain degree of import surveillance.

At the end of 2014 certain changes were being discussed to bring Regulation 258/2010 in line with other currently applicable emergency measures. These elements will result in an amendment of the measure in the beginning of 2015.

3.5.3.2 Requirements

Regulation 258/2010 regulates importation of guar gum intended for animal or human consumption and originating or consigned from India. It covers guar gum imported as such, but also compound feed and food in case this additive constitutes at least 10 % of the product (Article 1).

The following documents must accompany each consignment intended for EU import (Article 2):

1. A health certificate, signed by an authorised representative of the Ministry of Commerce and Industry of India, with a 4-month validity from the date of its issue: this document certifies that the product does not contain more than 0.01 mg/kg of PCP and
2. An analytical report, issued by an accredited laboratory of the exporting country: this document indicates the results of sampling and analysis for the presence of PCP, including the uncertainty of the analytical result and limits of detection and quantification of the analytical method.

Furthermore, business operators are required to give prior notification to one of the control points (CP), indicating the estimated date and time of arrival of the consignment (Article 4). Designation of CPs lies with Member States, which must draw up a national list for this purpose and make it available to the public and the European Commission.

With relation to official controls, the regular framework—documentary, identity and physical checks, including sampling and analysis—is required from Member

States' competent authorities. Identity and physical checks must be carried out with an intensity of at least 5 % of all arriving consignments. Whenever physical checks are performed, consignments must be kept under official supervision for a maximum of 15 working days, pending the availability of the results of the laboratory analysis. In addition to consignments of guar gum from India, Member States' competent authorities are also expected to perform random physical checks on consignments of the same product consigned from other non-EU countries (Article 5).

The Regulation explicitly prohibits any product found to contain more than 0.01 mg/kg PCP to enter the feed and food chain. Non-compliant products have to be safely disposed of in accordance with Article 19 of Regulation 882/2004 (Article 9).

3.5.4 Microbial Contamination in Imported Food of Non-animal Origin

3.5.4.1 *E. coli* Outbreak in Sprouts and Seeds for Sprouting from All Non-EU Counties (2011): Regulation (EU) No. 211/2013

From May to July 2011, a major outbreak occurred in the EU (Sect. 2.3.4) due to the contamination of sprouts by Shiga toxin-producing *Escherichia coli* (STEC) O104:H4 with clusters reported in Germany and France (European Commission 2011g). These microorganisms are highly pathogenic, and very low levels (i.e. as little as four bacteria/kg) in seeds for sprouting can be sufficient to cause outbreaks (EFSA 2011). Only in Germany, the outbreak concerned thousands of people, including over 50 fatalities (Robert Koch Institute 2011).

The European Commission coordinated the response to the outbreak at EU level and was assisted in this task by the European Centre for Disease Prevention and Control, EFSA and the EU Reference Laboratory for *E. coli*.

However, since the products that caused the outbreak could not be immediately identified, uncertainty and, to some extent, misinformation adversely affected the EU fruit and vegetables sector. Indeed, consumers refrained from buying tomatoes, cucumbers, lettuce, courgettes and sweet peppers, i.e. all those vegetables that, initially, had been considered at the origin of the outbreak. Ensuing losses for EU farmers, only in the first 2 weeks following the outbreak, were estimated at least in 812 million euros.

Investigations ultimately traced back the cause of the outbreak to a consignment of fenugreek seeds for sprouting from Egypt. In July 2011, the European Commission thus adopted emergency measures (Sect. 2.3.4) under Decision 2011/402/EU requiring EU Member States to withdraw from the market lots of seeds and other beans imported from that country into the EU in the period 2010–2011 (European Commission 2011a). At the same time, the measures foresaw a temporary suspension on the introduction of new imports into the EU, which was, subsequently, extended until March 2012 (European Commission 2011f). In the

Table 3.3 EU legislation adopted as a result of the *E. coli* crisis

Title of the measure	Scope	Applicability
Commission Implementing Regulation (EU) No. 208/2013	Traceability requirements for sprouts and seeds intended for the production of sprouts (excluding sprouts that have undergone a treatment which eliminates microbiological hazards)	1 July 2013
Commission Regulation (EU) No. 209/2013	Amending Regulation (EC) No. 2073/2005 as regards microbiological criteria for sprouts and the sampling rules for poultry carcases and fresh poultry meat	1 July 2013
Commission Regulation (EU) No. 210/2013	Approval of establishments producing sprouts pursuant to Regulation (EC) No. 852/2004 of the European Parliament and of the Council	1 July 2013
Commission Regulation (EU) No. 211/2013 (as amended by Commission Regulation (EC) No. 704/2014)	Certification requirements for imports into the EU of sprouts and seeds intended for the production of sprouts (derogation from this requirement by microbiological testing before export to the EU, until 1 July 2015)	1 April 2013

meantime, different actions have been performed by the European Commission[4] which eventually led to a new specific legislative framework, based on (Table 3.3):

(a) Commission Implementing Regulation (EU) No. 208/2013 (European Commission 2013b) with relation to traceability requirements
(b) Commission Regulation (EU) No. 209/2013 (European Commission 2013c) concerning microbiological criteria and sampling rules
(c) Commission Regulation (EU) No. 210/2013 (European Commission 2013d) with concern to approved establishments
(d) Commission Regulation (EU) No. 211/2013 (European Commission 2013e) concerning certification requirements for imports into the EU.

Since the entry into application of the new legislative framework, EU Member States have reported no RASFF notification.

---

[4] The FVO performed an audit in Egypt from 21 to 25 August 2011. Overall, the final DG (SANCO) 2011-6265 document has concluded that the presence of polluted waters, human populations and cattle in the proximity of the areas where the concerned products came from all posed a potential risk of contamination. The actual source of the contamination, however, could not be identified. In the follow-up of the outbreak, the Commission has also launched a media campaign in all EU Member States, reaching out an estimated number of 37 million people across the EU. 227 million euros were also earmarked as a market intervention to support those sectors within the EU that were most hit financially by the outbreak.

Requirements

Commission Regulation 211/2013 (as amended by Commission Regulation (EU) No. 704/2014) applies to sprouts or seeds for sprouting intended for EU import, with the exemption of sprouts that have undergone a treatment which eliminates microbiological hazards in line with EU requirements (Article 1). Based on Article 48 of Regulation 882/2004, the measure concerns all non-EU countries exporting those products to the EU.

Certification Requirements

Consignments of products covered by this measure must be accompanied by the certificate provided in the Annex to Regulation 704/2014 (amending 211/2013). The certificate must attest that the sprouts or seeds have been produced under conditions which comply with the general hygiene provisions for primary production and associated operations set out in Part A of Annex I to Regulation 852/2004 (Article 3). The certificate also states that the sprouts have been produced under conditions which comply with the traceability requirements laid down in the Regulation 208/2013, in establishments approved in accordance with the requirements laid down in Article 2 of the Regulation 210/2013 and respect the microbiological criteria laid down in Annex I to the Regulation 2073/2005.

However, recent FVO audits signalled that certain competent authorities in certain non-EU countries are not always capable of fulfilling the above-referred certification requirements (DG SANCO 2013c). For this reason, and while waiting for the corrective actions implemented by these countries (i.e. building a robust certification system), Regulation 704/2014 introduced the possibility of replacing the certificate by a microbiological test.[5]

3.5.4.2   Betel Leaves from Bangladesh: Decision 2014/88/EU (as Amended by Decision 2014/510/EU)

Safety of imports of betel leaves from Bangladesh is a long-standing issue. It emerged for the first time in 2011 as a result of several RASFF notifications signalling rejections at EU borders of these imports for contamination with different pathogenic strains of *Salmonella* and *E. coli*.

Betel leaves (also known as 'paan leaves' or 'betel quid') are traditional products intended as mild stimulant and breath freshener. They are generally exported

---

[5] With concern to seeds intended for the production of sprouts, the certification can be replaced by a microbiological test performed before export to the EU. This derogation is applicable only until 1 July 2015.

to EU countries where communities of Asian emigrants reside, including UK, Germany and, to a lesser extent, Italy. Despite being a niche product, in 2012, exports of betel leaves only from Bangladesh accounted for over 40 million euros.

Since 2011, EU Member States have notified through the RASFF over 140 cases of microbiological contamination in these products. Despite the high number of non-compliant consignments detected, the European Commission conceded time to the competent authorities in Bangladesh so that safety concerns associated with their exports could be addressed. In response, Bangladeshi authorities put a national action plan in place in order to ensure a pathogen-free production chain and suspended all exports of those products in November 2012.

However, in February 2013, the FVO audited the country (DG SANCO 2013b) and found that the national authorities had not yet fully implemented the action plan as well as that the export surveillance system for food of non-animal origin presented significant deficiencies (namely the lack of appropriate facilities and equipment for performing official controls prior to exports). In addition to that, in spite of a national ban on exports being in place, consignments of betel leaves were still arriving at EU borders and often reported as non-compliant.

Due to these developments, the European Commission has introduced a temporary import ban on the products in question by Decision 2014/88/EU (European Commission 2014d). The ban has recently been prolonged by Decision 2014/510/EU (European Commission 2014e) since there could be no significant improvements noted in the situation.

Finally, it is worth mentioning that, over the same period, Member States have reported several further RASFF notifications regarding betel leaves from other non-EU countries, namely India and Thailand. As a result, the Commission has included these imports (European Commission 2014f) in the list of feed and food subject to reinforced border controls under Regulation 669/2009. Ideally, increased border surveillance on imports of betel leaves from these other Asian countries will minimise opportunities for trade triangulation to get around EU import restrictions.

Requirements

Decision 2014/88/EU regulates the importation of consignments containing, or consisting of, betel leaves from Bangladesh. The Decision specifies that consignments to be controlled include those bearing a specific 'Combined Nomenclature' (CN) code; however, it also implies the spectrum of products subject to controls being broader, e.g. possibly including cases where betel leaves or products containing them are declared under other CN codes (Article 1).

Suspension of Import

The suspension of import of betel leaves from Bangladesh applies temporarily until 30 June 2015. However, the Decision is likely to be reviewed again in the light of the assurances that the exporting country may provide in the meantime.

3.6   Emergency Measures of Broader Scope Covering Imports of Non-animal Origin

3.6.1   Fukushima: Regulation (EU) No. 322/2014

3.6.1.1   Background

The incident at the *Fukushima Dai-ichi* power plant in 2011 has been a disaster of catastrophic proportions (Sect. 1.4.2). After the incident, substantial amounts of radioactive materials have been released progressively into the environment. As a consequence, this event has been considered the largest nuclear incident since the *Chernobyl* disaster in 1986 (World Nuclear Association 2014).

Shortly after the incident, the European Commission had been informed that radionuclide levels in certain food products originating from the region of *Fukushima* exceeded the maximum levels permitted by Japanese law. Consequently, the European Commission has immediately adopted EU emergency measures that became subject to regular review in order to take into account the evolution of the situation at any growing season. Table 3.4 provides a detailed chronology of the main reviews with concern to EU measures in question.

At present, after 3 years of strict surveillance prior to export at EU borders, the new Regulation (EU) No. 322/2014 has lightened import control requirements (European Commission 2014g) on the basis of satisfactory levels of compliance

Table 3.4   Chronology of EU emergency measures related to *Fukushima* nuclear power station incident

Title of emergency measure	Adoption date	Review/ applicability	Status
Commission Implementing Regulation (EU) No. 297/2011 (European Commission 2011b), followed by amendments	25 March 2011	Reviewed monthly	First legislation following nuclear accident
Commission Implementing Regulation (EU) No. 961/2011 (European Commission 2011c), followed by amendments	27 September 2011	Reviewed monthly	Repealing 297/2011
Commission Implementing Regulation (EU) No. 284/2012 (European Commission 2012b), followed by amendments	29 March 2012	Reviewed regularly	Repealing 961/2011
Commission Implementing Regulation (EU) No. 996/2012 (European Commission 2012c), followed by amendments	26 October 2012	Applicable until 31 March 2014	Repealing 284/2012
Commission Implementing Regulation (EU) No. 322/2014	28 March 2014	To be reviewed before 31 March 2015	Legislation currently in force

by concerned exports from Japan. New actions should apply until the next growing season (March 2015) when the situation will be further reviewed.

3.6.1.2 Requirements

The scope of Regulation 322/2014 is broad and complex. It covers feed and food imports into the EU market from Japan with the following exceptions (Article 1):

- Products which left Japan before 28 March 2011
- Products which have been harvested and/or processed before 11 March 2011
- Certain alcoholic beverages
- Certain personal consignments.

Products destined for the EU import must comply with the maximum levels of caesium-134 (Cs-134) and caesium-137 (Cs-137) as set in Annex II and Annex III to the Regulation (Article 4).

Declaration and Analytical Report

Each consignment of products (with the exception of certain tea originating from prefectures other than *Fukushima*) must be accompanied with a valid declaration drawn up and signed by the Japanese competent authorities or an authorised representative (Articles 5 and 6). As appropriate, the declaration should serve for the following:

1. Certify that the products comply with the legislation in force in Japan
2. Specify whether products are falling or not under the transitional measures provided for in the Japanese legislation (Article 16);
3. Certify that the product has been harvested and/or processed before 11 March 2011; or
4. Indicate that a certain product originates and/or is consigned from a given prefecture (Article 5).

Moreover, exports originating from prefectures particularly affected by radioactivity (e.g. most products from *Fukushima*, mushroom from *Nagano* or fish from *Tochigi* and others listed in Annex IV to the Regulation) require, in addition to the declaration, an analytical report containing the results of sampling and analysis carried out by the Japanese authorities.

Obligations for Feed and Food Business Operators

Imports of non-animal origin must be presented at DPE as defined by Article 3 letter b of Regulation 669/2009, while imports of animal origin need to be channelled through Border Inspection Posts (BIP). For this reason, business operators must give prior notification to the competent authorities of the DPE/BIP of the arrival of consignments covered by Regulation 322/2014. Notification must be given at least two working days before the envisaged arrival by filling the relevant part of the CED (Article 9).

Performance of Official Controls and Release for Free Circulation

With reference to official controls, competent authorities of DPE (or BIP, in case of import of animal origin) must perform:

1. Documentary checks on all consignments of products and on the ones that must be accompanied by the declaration (Article 5, paragraph 4)
2. Random identity and physical checks, including laboratory analysis in the presence of Cs-134 and Cs-137.

When laboratory tests are performed, the analytical results must be available within a maximum of five working days. In case results of the laboratory analysis reveal that the content of the declaration accompanying the consignment is false, the declaration is considered invalid and the consignment not compliant with the Regulation (Article 10). Non-compliant products must be disposed of or returned to Japan (Article 13).

3.6.2 GMO Rice and Rice Products from China Decision 2011/884/EU (as Amended by 2013/287/EU)

3.6.2.1 Background

In September 2006, some EU Member States reported through the RASFF several cases of Chinese rice products contaminated with the (unauthorised) genetically modified (GMO) rice type Bt 63. In response to that, the European Commission has adopted emergency measures under Decision 2008/289/EC (European Commission 2011d), with the objective to prevent further importation into the EU of similar products.

In 2010, the presence of new rice varieties containing other unauthorised GMO types (*Kefeng* 6 and *Kemingdao* 1) was reported. Furthermore, two subsequent audits performed by the FVO in China in 2008 and 2011 (DG SANCO 2008, 2011c) have confirmed the presence of a high risk of further introductions into the EU of rice products contaminated with unauthorised GMO.

This ultimately resulted in the broadening of the range of products and the strengthening of the import requirements applicable through the adoption of Decision 2011/884/EU (European Commission 2011e). Following the entry into force of this Decision, the number of RASFF notifications ensuing from official controls increased significantly. Moreover, experience gained by Member States in implementing the requirements of the decision signalled the need to further widen the spectrum of products subject to measures and to adjust certain aspects of the sampling protocol. This ultimately led to the adoption of Decision 2013/287/EU (European Commission 2013f), which, applicable as of 4 July 2013, amends Decision 2011/884/EU.

3.6.2.2 Requirements

Decision 2013/287/EU applies to a number of feed and food products listed in its Annex I originating in or consigned from China (Article 1). The scope covers both rice products of non-animal origin (e.g. rice flour) and animal origin (e.g. uncooked pasta containing eggs). It foresees that import requirements must be reviewed regularly so as to take into account new developments in the situation, including scientific and technical progress in the methods for sampling and analysis (Article 10).

Obligations of Feed and Food Business Operators

Business operators exporting products of non-animal origin covered by the emergency measure in place must give adequate prior notification of the estimated date and time of the physical arrival of the consignment and of its nature to the competent authorities of the relevant DPE. To this end, they must use and complete the CED. The notification to the DPE must be made at least one working day prior to the physical arrival of the consignment. In case of import of animal origin, the Common Veterinary Entry Document (CVED), as provided for in Regulation (EC) No. 136/2004 (European Commission 2004c) must be transmitted to the BIP (Article 3).

Each consignment must be accompanied by an analytical report for each lot and a health certificate in accordance with the models set out in the other Annexes to the Decision. Both documents must be signed and verified by an authorised representative of the Chinese competent authorities. Where a product is not containing, consisting of or produced from rice, the analytical report and the health certificate may be replaced by a statement from the business operator indicating that the feed or food is not containing, consisting of or produced from rice.

Additionally, each consignment must bear an identification code which should appear on all documents accompanying the consignment and on each individual bag of which the consignment consists of (Article 4, paragraph 4).

Performance of Official Controls and Release for Free Circulation

In order to ensure that the import conditions are complied with, each consignment of products must be subject to systematic (i.e. 100 %) documentary checks by competent authorities at the DPE (or BIP if the product is of animal origin).

If the health certificate or the analytical report is missing, the consignment must be redispatched to China or, alternatively, destroyed, in accordance with Article 5, paragraph 2.

After the documentary checks, the competent authority of the DPE/BIP must perform systematic (i.e. 100 %) physical checks, including sampling and analysis for the presence of unauthorised GMO. If the consignment consists of several lots, sampling and analysis must be performed on each lot. The competent authority of the DPE may allow onward transportation of the consignment until its final destination, while results of the controls are still pending, provided that the consignment remains under the supervision of the competent authorities. The release for free circulation is only possible where the results of the control activities performed indicate full compliance with the applicable EU requirements (Article 5, paragraphs 3, 4 and 5).

3.6.3 Melamine in Milk and Soy from China: Regulation (EU) No. 1135/2009

3.6.3.1 Background

In September 2008, high levels of melamine were found in infant milk and other milk products with Chinese origin. In order to counter the health risk that could result from the exposure to high melamine content in feed and food products, the European Commission adopted Decision 2008/798/EC (European Commission 2008b) in autumn 2008. This measure prohibited the importation into the EU any product containing milk or milk products, and soya or soya products intended for the particular nutritional use of infants and young children. It also required EU Member States to perform systematic (i.e. 100 %) controls on all consignments originating in or consigned from China of feed and food containing milk, milk products, soya or soya products and of ammonium bicarbonate intended for feed and food.

Following the adoption of Decision 2008/798/EC, the number of RASFF notifications concerning melamine in Chinese products decreased considerably. Chinese authorities had also implemented strict surveillance on their exports prior to dispatch. Against this background, Decision 2008/798/EC was reviewed in order to relax checks at import. Nevertheless, with reference to products intended for infants and young children, because of the role they play in the nourishment of those individuals, the Commission decided to maintain the import prohibition. The withrawal of this measure is however currently being considered.

Requirements

Regulation (EC) No. 1135/2009 (European Commission 2009d) foresees special import restrictions for feed and food containing milk, milk products, soya or soya products and ammonium bicarbonate from China.

Suspension of Import

At present, the import of products containing milk, milk products, soya or soya products intended for the particular nutritional use of infants and young children is prohibited into the EU. Member States must ensure that none of such products is present on the market and, in case is found, is immediately withdrawn and destroyed (Article 2).

Obligations of Feed and Food Business Operators

As concerns ammonium bicarbonate intended for feed and food or feed and food containing milk, milk products, soya or soya products from China, business operators must give prior notification to the relevant CP of the estimated date and time of arrival of the consignment (Article 3).

Performance of Official Controls and Release for Free Circulation

EU Member States are required to carry out systematic (i.e. 100 %) documentary checks on incoming consignments of products subject to prior notification and approximately 20 % identity and physical checks, including sampling and laboratory analysis in order to verify possible presence of melamine. Consignments must be kept under official supervision, pending the availability of the results of the laboratory analysis (i.e. no possibility for 'onward transportation') (Article 4, paragraph 1).

Additionally, EU Member States may carry out random physical checks on other feed and food products with a high protein content originating from China. In any event, if a product is found to contain more than 2.5 mg/kg melamine, it must not be allowed to enter the feed and food chain and must be safely disposed, in accordance with Article 4, paragraph 2.

All official controls must be carried out at designated CP for this purpose. EU Member States must communicate their CP to the European Commission, in accordance with Article 4, paragraph 3 (Table 3.5).

Table 3.5 Summary of EU emergency measures

No.	EU legislation	Legal basis	Date of applicability	Scope (products/country of origin covered)	Hazard	Prior notification (including use of CED)	Import conditions	Import controls	Review clause
1	Commission Implementing Regulation (EU) No. 884/2014	Art. 53(1)(b)(ii) of Reg. (EC) No. 178/2002	20th day following publication in the OJ	Aflatoxin	Brazil nuts from Brazil; groundnuts from China, Egypt, Ghana and India; pistachios from Iran; dried figs, hazelnuts, pistachios, etc., from Turkey; watermelon seeds from Nigeria	Yes, via CED sent to DPE (and DPI) (at least 1 working day prior to physical arrival)	Results of sampling and analysis performed by the country of origin; health certificate (valid for 4 months)	Place of control: DPE (and DPI) Intensity of control: Documentary check: 100 %; physical check: depending on the commodity, as indicated in Annex I (random, 20 or 50 %)	–
2	Commission Implementing Regulation (EU) No. 885/2014	Art. 53(1)(b)(ii) of Reg. (EC) No. 178/2002	3rd day following publication in the OJ	Pesticide residue	Okra (food, fresh and frozen) and curry leaves (food, herb) from India	Yes, via CED sent to DPE (at least 1 working day prior to physical arrival)	Results of sampling and analysis performed by the country of origin; health certificate (valid for 4 months)	Place of control: DPE Intensity of control: documentary check: 100 %; physical check: 20 %	–

(continued)

Table 3.5 (continued)

No.	EU legislation	Legal basis	Date of applicability	Scope (products/country of origin covered)	Hazard	Prior notification (including use of CED)	Import conditions	Import controls	Review clause
3	Regulation (EU) No. 258/2010	Art. 53(1)(b)(ii) of Reg. (EC) No. 178/2002	20th day following publication in the OJ	Guar gum or products containing at least 10 % guar gum from India	Pentachlorophenol (PCP) and dioxins	Yes, sent to control point	Health certificate (valid for 4 months); Analytical report issued by an accredited laboratory on a sample taken by the Indian CA	Place of control: at control points specifically designated by the Member States for that purpose Intensity of control: documentary check: 100 %; physical check: at least 5 % from India and random physical checks from countries other than India	–

(continued)

Table 3.5 (continued)

No.	EU legislation	Legal basis	Date of applicability	Scope (products/country of origin covered)	Hazard	Prior notification (including use of CED)	Import conditions	Import controls	Review clause
4	Commission Regulation (EU) No. 211/2013	Art. 48(1) of Reg.(EC) No. 882/2004	20th day following publication in the OJ	Sprouts of seeds intended for the production of sprouts from all non-EU countries	Shiga toxin-producing *E. coli* (STEC) (bacterial pathogens)	–	Import (health) certificate attesting compliance with general hygiene, traceability, approved establishment and microbiological provisions	–	–
5	Commission Implementing Decision 2014/88/EU	Art. 53(1)(b)(i) of Reg. (EC) No. 178/2002	Immediate, as of 13 February 2014 until at least 31 July 2014	Betel leaves ('Piper betle') from Bangladesh	*Salmonella*, E. coli	–	Temporary suspension of import (full ban)	–	Yes, after 31 July 2014

(continued)

Table 3.5 (continued)

No.	EU legislation	Legal basis	Date of applicability	Scope (products/country of origin covered)	Hazard	Prior notification (including use of CED)	Import conditions	Import controls	Review clause
6	Commission Implementing Regulation 322/2014	Art. 53(1)(b)(ii) of Reg. (EC) No. 178/2002	1 April 2014	All feed and food (with certain exclusion) from Japan	Radioactive contamination (caesium)	Yes, with the exception of tea originating from prefectures other than Fukushima Prior notification via CED is sent to DPE (for import of non-animal origin) and to BIP (for import of animal origin) at least 2 working days prior to physical arrival	Declaration (with the exception of tea originating from prefectures other than Fukushima) signed by an authorised representative of the competent Japanese authorities; analytical report only in case of certain (detailed) consignments	Place of controls: DPE (BIP) Intensity of control: documentary check: 100 %; physical check: random frequency	Yes, by 31 March 2015

(continued)

Table 3.5 (continued)

No.	EU legislation	Legal basis	Date of applicability	Scope (products/country of origin covered)	Hazard	Prior notification (including use of CED)	Import conditions	Import controls	Review clause
7	Commission Decision 2011/884/EU (amended by 2013/287/EU)	Art. 53(1) of Reg. (EC) No. 178/2002	20th day following publication in the OJ	Various rice products listed in its annex	Non-authorised GMO	Yes, via CED (CVED) sent to DPE (BIP) at least 1 working day prior to physical arrival	Analytical report and health certificate completed, signed and verified by an authorised representative of the 'Entry Exit Inspection and Quarantine Bureau of the People's Republic of China' (AQSIQ) (or statement from the operator indicating that the food or feed does not contain, consist of or is produced from rice)	Place of controls: DPE (BIP) Intensity of control: documentary check: 100 %; physical check: 100 % (in case accompanied by analytical report and health certificate)	Yes, regular review to take into account new developments

(continued)

Table 3.5 (continued)

No.	EU legislation	Legal basis	Date of applicability	Scope (products/country of origin covered)	Hazard	Prior notification (including use of CED)	Import conditions	Import controls	Review clause
8	Regulation (EU) No. 1135/2009	Art. 53(1)(b) of Reg. (EC) No. 178/2002	20th day following publication in the OJ	Feed and food containing milk, milk products, soya or soya products (including foodstuffs intended for particular nutritional uses); ammonium bicarbonate intended for food and feed from China	Melamine	Yes, to control point (specifically designated by EU Member States)	Import prohibition (full ban) for infant nutrition purposes	Place of controls for ammonium bicarbonate intended for food and feed and of food and feed containing milk (products) or soya (products): specifically designated control point. Intensity of control: documentary check: 100 %; physical check: approximately 20 %. Random physical checks on other feed and food products with a high protein content originating from China	–

References

DG SANCO (2007) Final report of a mission carried out in India from 05 to 11 October 2007 to gather information on the source of contamination of guar gum with pentachlorophenol and dioxins and to assess the control measures put in place by the Indian authorities to avoid a reoccurrence of this contamination. DG (SANCO) 2007-7619—MR final. Commission of the European Communities, Brussels. Available: http://ec.europa.eu/food/fvo/act_getPDF.cfm?PDF_ID=6406. Accessed 28 Aug 2014

DG SANCO (2008) Final Report of a mission carried out in China from 27 November to 03 December 2008 in order to evaluate official control systems for Commission Decision 2008/289/EC on emergency measures regarding the unauthorised GMO "Bt 63" in rice products. DG (SANCO)/2008-7834 Final. European Commission, Brussels. Available: http://ec.europa.eu/food/fvo/act_getPDF.cfm?PDF_ID=7450. Accessed 28 Aug 2014

DG SANCO (2009) Final report of a mission carried out in India from 01 to 12 October 2009 in order to assess the control measures put in place by the Indian authorities to prevent contamination of guar gum with pentachlorophenol (PCP) and dioxins and to follow-up the recommendations of mission DG SANCO 2007-7619. DG (SANCO) 2009-8329—MR Final European Commission, Brussels. Available: http://ec.europa.eu/food/fvo/act_getPDF.cfm?PDF_ID=8022. Accessed 28 Aug 2014

DG SANCO (2010) Overview report (FVO) on the outcome of a mission series carried out in Third Countries on contaminants (mycotoxins, PCP, dioxins and 3-mcpd) from May 2001 to March 2010, OR No. 2010-8965. European Commission, Brussels. Available: http://ec.europa.eu/food/fvo/specialreports/2010_8965_or_contaminants.pdf. Accessed 28 Aug 2014

DG SANCO (2011a) Final report of an audit carried out from 30 November to 08 December 2011 in India in order to evaluate controls of pesticides in food of plant origin intended for export to the European Union. DG (SANCO) 2011-6037—MR Final. European Commission, Brussels. Available: http://ec.europa.eu/food/fvo/act_getPDF.cfm?PDF_ID=9611. Accessed 28 Aug 2014

DG SANCO (2011b) Final report of an audit carried out from 18 to 24 October 2011 in India in order to assess the controls on guar gum for export to the European Union. DG (SANCO) 2011-6060—MR Final. European Commission, Brussels. Available: http://ec.europa.eu/food/fvo/act_getPDF.cfm?PDF_ID=9445. Accessed 28 Aug 2014

DG SANCO (2011c) Final report of an audit carried out in China from 29 March to 08 April 2011 in order to evaluate the control systems for genetically modified organisms (GMOs) in respect of seed, food and feed intended for export to the EU. DG (SANCO)/2011-6208 Final. European Commission, Brussels. Available: http://ec.europa.eu/food/fvo/act_getPDF.cfm?PDF_ID=9269. Accessed 28 Aug 2014

DG SANCO (2011d) Final report of an audit carried out in Brazil from 02 to 12 May 2011 with reference to the evaluation of controls of aflatoxin contamination in peanuts intended for export into the EU. DG (SANCO) 2011-6034—MR Final report is available at: http://ec.europa.eu/food/fvo/act_getPDF.cfm?PDF_ID=9149

DG SANCO (2011e) Final report of an audit carried out in China from 20 to 29 September 2011 with reference to the assessment of the controls of Aflatoxin contamination for export to the European Union. The DG (SANCO) 2011-6038MR Final report is available at this web address—MR Final report is available at this web address: http://ec.europa.eu/food/fvo/rep_details_en.cfm?rep_id=2818

DG SANCO (2012) Final report of an audit carried out in Turkey from 09 to 16 October 2012 with relation to the assessment of the control system of aflatoxin contamination in hazelnuts and dried figs. DG (SANCO) 2012-6292—MR Final report is available at: http://ec.europa.eu/food/fvo/act_getPDF.cfm?PDF_ID=10210

DG SANCO (2013a) Final report of an audit carried out from 21 October to 01 November 2013 in India in order to assess the controls of aflatoxin contamination in peanuts intended for export into the European Union and follow up mission. DG (SANCO) 2013-6683—MR Final. European Commission, Brussels. Available: http://ec.europa.eu/food/fvo/act_getPDF.cfm?PDF_ID=10969. Accessed 28 Aug 2014

DG SANCO (2013b) Final report of an audit carried out in Bangladesh from 30 January to 07 February 2013 in order to evaluate the system of official controls for export of plants to the European Union, DG (SANCO) 2013-6815—MR Final. European Commission, Brussels. Available: http://ec.europa.eu/food/fvo/rep_details_en.cfm?rep_id=3082. Accessed 10 Dec 2014

DG SANCO (2013c) Final report of an audit carried out in China from 14 to 18 October 2013 with relation to the assessment of the control of microbiological contamination in seeds for human consumption, DG (SANCO) 2013-6680—MR Final. European Commission, Brussels. Available: http://ec.europa.eu/food/fvo/rep_details_en.cfm?rep_id=3241

EFSA (2011) EFSA assesses the public health risk of seeds and sprouted seeds. Available: http://www.efsa.europa.eu/en/press/news/111115.htm. Accessed 28 Aug 2014

European Commission (2000) Commission Decision 2000/49/EC of 6 December 1999 repealing Decision 1999/356/EC and imposing special conditions on the import of peanuts and certain products derived from peanuts originating in or consigned from Egypt. Off J Eur Union L19:46–50

European Commission (2001) Commission Regulation (EC) No. 466/2001 of 8 March 2001 setting maximum levels for certain contaminants in foodstuffs. Off J Eur Union L77:1–13

European Commission (2002a) Commission Decision 2002/79/EC of 4 February 2002 imposing special conditions on the import of peanuts and certain products derived from peanuts originating in or consigned from China. Off J Eur Union L34:21–25

European Commission (2002b) Commission Decision 2002/80/EC of 4 February 2002 imposing special conditions on the import of figs, hazelnuts and pistachios and certain products derived thereof originating in or consigned from Turkey. Off J Eur Union L34:26–30

European Commission (2003) Commission Decision 2003/493/EC of 4 July 2003 imposing special conditions on the import of Brazil nuts in shell originating in or consigned from Brazil. Off J Eur Union L168:33–38

European Commission (2004a) Commission Decision 2004/429/EC amending Decisions 97/830/EC, 2000/49/EC, 2002/79/EC and 2002/80/EC as regards the points of entry through which the products concerned may only be imported into the community. Off J Eur Union L154:20–35

European Commission (2004b) Commission Decision 2004/428/EC of 29 April 2004 amending Decision 2003/493/EC as regards the points of entry through which Brazil nuts in shell originating in or consigned from Brazil may only be imported into the Community. Off J Eur Union L154:14–19

European Commission (2004c) Commission Regulation (EC) No. 136/2004 of 22 January 2004 laying down procedures for veterinary checks at Community Border Inspection Posts on products imported from third countries. Off J Eur Union L21:11

European Commission (2005) Commission Decision 2005/85/EC of 26 January 2005 imposing special conditions on the import of pistachios and certain products derived from pistachios originating in or consigned from Iran. Off J Eur Union L30:12–18

European Commission (2006a) Commission Decision 2006/504/EC of 12 July 2006 on special conditions governing certain foodstuffs imported from certain third countries due to contamination risks of these products by aflatoxins. Off J Eur Union L199:21–32

European Commission (2006b) Commission Regulation (EC) No. 1881/2006 of 19 December 2006 setting maximum levels for certain contaminants in foodstuffs. Off J Eur Union L364:5–24

European Commission (2006c) Commission Regulation (EC) No. 401/2006 of 23 February 2006 laying down the methods of sampling and analysis for the official control of the levels of mycotoxins in foodstuffs. Off J Eur Union L70:12–34

European Commission (2008a) Commission Decision 2008/352/EC of 29 April 2008 imposing special conditions governing guar gum originating in or consigned from India due to contamination risks of those products by pentachlorophenol and dioxins. Off J Eur Union L117:42–44

European Commission (2008b) Commission Decision 2008/798/EC of 14 October 2008 imposing special conditions governing the import of products containing milk or milk products originating in or consigned from China, and repealing Commission Decision 2008/757/EC. Off J Eur Union L273:18–20

European Commission (2009a) Commission Regulation (EC) No. 669/2009 of 24 July 2009 implementing Regulation (EC) No. 882/2004 of the European Parliament and of the Council as regards the increased level of official controls on imports of certain feed and food of non-animal origin and amending Decision 2006/504/EC. Off J L194:11–21

European Commission (2009b) Commission Regulation (EC) No. 1152/2009 of 27 November 2009 imposing special conditions governing the import of certain foodstuffs from certain third countries due to contamination risk by aflatoxins and repealing Decision 2006/504/EC. Off J Eur Union L313:40–49

European Commission (2009c) Commission Regulation (EC) No. 152/2009 of 27 January 2009 laying down the methods of sampling and analysis for the official control of feed. Off J Eur Union L54:1–130

European Commission (2009d) Commission Regulation (EC) No. 1135/2009 of 25 November 2009 imposing special conditions governing the import of certain products originating in or consigned from China, and repealing Commission Decision 2008/798/EC. Off J Eur Union L311:3–5

European Commission (2010) Commission Regulation (EC) No. 258/2010 of 25 March 2010 imposing special conditions on the imports of guar gum originating in or consigned from India due to contamination risks by pentachlorophenol and dioxins, and repealing Decision 2008/352/EC. Off J Eur Union L80:28–31

European Commission (2011a) Commission Implementing Decision 2011/402/EU of 6 July 2011 on emergency measures applicable to fenugreek seeds and certain seeds and beans imported from Egypt. Off J Eur Union L179:10–12

European Commission (2011b) Commission Implementing Regulation (EU) No. 297/2011 of 25 March 2011 imposing special conditions governing the import of feed and food originating in or consigned from Japan following the accident at the Fukushima nuclear power station. Off J Eur Union L80:5–8

European Commission (2011c) Commission Implementing Regulation (EU) No. 961/2011 of 27 September 2011 imposing special conditions governing the import of feed and food originating in or consigned from Japan following the accident at the Fukushima nuclear power station and repealing Regulation (EU) No. 297/2011. Off J Eur Union L252:10–15

European Commission (2011d) Commission Implementing Decision 2008/289/EC of 22 December 2011 on emergency measures regarding unauthorised genetically modified rice in rice products originating from China and repealing Decision 2008/289/EC. Off J Eur Union L343:140–148

European Commission (2011e) Commission Decision 2011/884/EU of 22 December 2011 on emergency measures regarding unauthorised genetically modified rice in rice products originating from China and repealing Decision 2008/289/EC. Off J Eur Union L343:140–148

European Commission (2011f) Commission Implementing Decision 2011/880/EU of 21 December 2011 amending Annex I to Implementing Decision 2011/402/EU on emergency measures applicable to fenugreek seeds and certain seeds and beans imported from Egypt. Off J Eur Union L343:117–118

European Commission (2011g) Lessons learned from the 2011 outbreak of Shiga toxin-producing Escherichia coli (STEC) O104:H4 in sprouted seeds. Commission Staff working document SANCO 13004/2011. Commission of the European Communities, Brussels. Available: http://ec.europa.eu/food/food/biosafety/salmonella/docs/cswd_lessons_learned_en.pdf. Accessed 28 Aug 2014

European Commission (2012a) Commission Implementing Regulation (EU) No. 274/2012 of 27 March 2012 amending Regulation (EC) No. 1152/2009 imposing special conditions governing the import of certain foodstuffs from certain third countries due to contamination risk by aflatoxins. Off J Eur Union L90:14–16

European Commission (2012b) Commission Implementing Regulation (EU) No. 284/2012 of 29 March 2012 imposing special conditions governing the import of feed and food originating in or consigned from Japan following the accident at the Fukushima nuclear power station and repealing Implementing Regulation (EU) No. 961/2011. Off J Eur Union L92:16–23

European Commission (2012c) Commission Implementing Regulation (EU) No. 996/2012 of 26 October 2012 imposing special conditions governing the import of feed and food originating in or consigned from Japan following the accident at the Fukushima nuclear power station and repealing Implementing Regulation (EU) No. 284/2012. Off J Eur Union L299:31–41

European Commission (2013a) Commission Implementing Regulation (EU) No. 91/2013 of 31 January 2013 laying down specific conditions applicable to the import of groundnuts from Ghana and India, okra and curry leaves from India and watermelon seeds from Nigeria and amending Regulations (EC) 669/2009 and (EC) 1152/2009. Off J Eur Union L33:2–10

European Commission (2013b) Commission Implementing Regulation (EU) No. 208/2013 of 11 March 2013 on traceability requirements for sprouts and seeds intended for the production of sprouts. Off J Eur Union L68:16–18

European Commission (2013c) Commission Regulation (EU) No. 209/2013 of 11 March 2013 amending Regulation (EC) No. 2073/2005 as regards microbiological criteria for sprouts and the sampling rules for poultry carcases and fresh poultry meat. Off J Eur Union L68:19–23

European Commission (2013d) Commission Regulation (EU) No. 210/2013 of 11 March 2013 on the approval of establishments producing sprouts pursuant to Regulation (EC) No. 852/2004 of the European Parliament and of the Council. Off J Eur Union L68:24–25

European Commission (2013e) Commission Regulation (EU) No. 211/2013 of 11 March 2013 on certification requirements for imports into the Union of sprouts and seeds intended for the production of sprouts. Off J Eur Union L68:26–29

European Commission (2013f) Commission Implementing Decision 2013/287/EU of 13 June 2013 amending Implementing Decision 2011/884/EU on emergency measures regarding unauthorised genetically modified rice in rice products originating from China. Off J Eur Union L162:10–14

European Commission (2014a) Commission Implementing Regulation (EU) No. 884/2014 of 13 August 2014 imposing special conditions governing the import of certain feed and food from certain third countries due to contamination risk by aflatoxins and repealing Regulation (EC) No. 1152/2009. Off J Eur Union L242:4–19

European Commission (2014b) Commission Regulation (EU) No. 519/2014 of 16 May 2014 amending Regulation (EC) No. 401/2006 as regards methods of sampling of large lots, spices and food supplements, performance criteria for T-2, HT-2 toxin and citrinin and screening methods of analysis. Off J Eur Union L147:29–43

European Commission (2014c) Commission Implementing Regulation (EU) No. 885/2014 of 13 August 2014 laying down specific conditions applicable to the import of okra and curry leaves from India and repealing Implementing Regulation (EU) No. 91/2013. Off J Eur Union L242:20–26

European Commission (2014d) Commission Implementing Decision 2014/88/EU of 13 February 2014 suspending temporarily imports from Bangladesh of foodstuffs containing or consisting of betel leaves ('Piper betel'). Off J Eur Union L45:34–35

European Commission (2014e) Commission Implementing Decision 2014/510/EU of 31 July 2014 amending Implementing Decision 2014/88/EU suspending temporarily imports from Bangladesh of foodstuffs containing or consisting of betel leaves ('Piper betle') as regards its period of application. Off J Eur Union L228:33–34

European Commission (2014f) Commission Implementing Regulation (EU) No. 323/2014 of 28 March 2014 amending Annexes I and II to Regulation (EC) No. 669/2009 implementing Regulation (EC) No. 882/2004 of the European Parliament and of the Council as regards the increased level of official controls on imports of certain feed and food of non-animal origin. Off J Eur Union L95:12–23

European Commission (2014g) Commission Implementing Regulation (EU) No. 322/2014 of 28 March 2014 imposing special conditions governing the import of feed and food originating in or consigned from Japan following the accident at the Fukushima nuclear power station. Off J Eur Union L95:1–11

European Parliament and Council (2002) Regulation (EC) No 178/2002 of the European Parliament and of the Council of 28 January 2002 laying down the general principles and requirements of food law, establishing the European Food Safety Authority and laying down procedures in matters of food safety. Off J Eur Union L31:1–24

European Parliament and Council (2004) Regulation (EC) No. 882/2004 of the European Parliament and of the Council of 29 April 2004 on official controls performed to ensure the verification of compliance with feed and food law, animal health and welfare rules. Off J Eur Union L165:1–141

Robert Koch Institute (2011) Final presentation and evaluation of epidemiological findings in the EHEC O104:H4 outbreak, Germany 2011. Robert Koch Institute, Berlin. Available: http://www.rki.de/EN/Home/EHEC_final_report.pdf?__blob=publicationFile. Accessed 28 Aug 2014

World Nuclear Association (2014) Fukushima accident. World Nuclear Association. Available: http://www.world-nuclear.org/info/safety-and-security/safety-of-plants/fukushima-accident/. Accessed 28 Aug 2014

Chapter 4
Conclusion

Veronika Jezsó

Potential contamination of feed or food affects all countries and may often have a substantial impact on public health, animal health and the economy. On the other hand, consumers rightly expect and demand the highest level of safety when it comes to what they eat. For this reason, feed and food safety requires ongoing monitoring by all stakeholders, including market surveillance by competent authorities and international cooperation. Education and training play also a pivotal role in ensuring that the highest safety standards are met.

It is evident that feed and food safety is currently one of the main objectives of the EU, embracing all segments of the relevant chain, including primary production, processing and distribution. In this context, EFSA strives to strengthen public trust in the EU food safety system, by providing advice based on sound science and operating in a transparent, open and independent manner.

Historically, policy-makers, at EU and national level, have focussed their attention and efforts primarily on zoonoses, i.e. human diseases transmitted by animals or food of animal origin. However, over the last few years, a shift in this approach has been observed with policy-makers becoming increasingly aware that feed and food of non-animal origin are also potential carriers of contaminants and may, thus, involve serious health risks.

As regards imports of non-animal origin, recent food scares, such as the *E. coli* in 2011, and RASFF data confirm that a certain level of surveillance by the EU on feed and food in provenance from other countries is warranted.

All this considered, it does not come as a surprise that the amount of EU legislation regulating the safety of imports of non-animal origin has been growing significantly over the past years to the extent that it can be now regarded as a self-standing area of EU food law.

Indeed, the EU is currently entrusted with a wide range of policy tools in this area, including market surveillance, reinforced border controls, approval of pre-export checks and emergency measures: altogether they allow the adoption of the most appropriate risk management measures, or the modification of the existing ones, on a case-by-case basis.

© The Author(s) 2015
F. Montanari et al., *Risk Regulation in Non-Animal Food Imports*,
Chemistry of Foods, DOI 10.1007/978-3-319-14014-8_4

From what precedes, one could conclude that the current design of EU policy on imports of non-animal origin is fit for purpose insofar as it enables EU decision-makers to develop effective and proportionate responses to new as well as known risks.

If, on the one hand, based on the extensive analysis provided in this book, we would tend to agree with such conclusions, on the other, we believe that some improvements could be introduced in order to ensure greater consistency across this policy area.

More precisely, after 5 years of implementation, the EU regime of reinforced border surveillance provided for by Regulation (EC) No. 669/2009 appears to be well established and functioning to a satisfactory degree. Indeed, such a regime provides all stakeholders (i.e. non-EU countries, exporters/importers and national authorities at import) with a clear legal framework for official controls on certain imports of non-animal origin, based on the following:

1. Prenotification of consignments through a common document (i.e. CED);
2. Mandatory presentation at designated points of entry on the EU territory (i.e. DPEs);
3. Regular review of the list of imports subject to border surveillance, based on results of the checks performed by EU Member States; and
4. Publication of reports by the European Commission detailing results of official controls performed by EU Member States.

If one takes such a regime as a common baseline for the whole area governing imports of non-animal origin, he/she will immediately notice that not all the emergency measures introduced or amended following the adoption of Regulation 669/2009 are wholly consistent with the provisions of this latter.

To exemplify, Regulation (EU) No. 258/2010 setting import conditions for guar gum from India foresees prenotification of relevant consignments, although not through submission of a CED, and mandatory channelling through CPs on the EU territory (instead of DPEs). The same consideration applies to Regulation (EC) No. 1135/2009 regarding feed and food with Chinese origin at risk of contamination with melamine, where, again, a CED is not used and business operators must present their consignments at CPs designated for this purpose.

Furthermore, Regulation (EU) No. 884/2014 still maintains a control regime for products presenting a high risk of aflatoxin contamination involving DPEs and DPIs at the same time, a control regime that is different from the one designed by Regulation 669/2009 for similar product/risk combination although of a lower risk. It is a fact that exporters, importers as well as enforcement authorities still do not fully understand the reasons behind the coexistence of two different regimes within the same area.

Also, with a few exceptions (i.e. Fukushima's import restrictions and the EU ban on Bangladeshi paan leaves, Sects. 3.5.4.2 and 3.6.1), none of the emergency measures currently in place explicitly includes a review clause requiring the European Commission to reassess the situation, at pre-established dates, in the light of the prevalence of the risk and/or based on the results of official controls. In addition to that, Regulation 669/2009 remains the only piece of legislation across the area of imports of non-animal origin where full publicity is given to results of the official controls carried out by national authorities in EU Member States.

Based on the above, we are of the view that increased uniformity in the terminology used and in the modalities for performing official controls, together with full transparency when it comes to the outcome of said controls, is an essential step to ensure the current EU policy approach to imports of non-animal origin is fully consistent. Whereas increased uniformity would ultimately result in greater ease of implementation of the applicable import requirements by competent authorities of EU Member States, full transparency would, instead, provide further confirmation, in EU trade partners' eyes, that such requirements are evidence-based and remain proportionate over time.

CPSIA information can be obtained at www.ICGtesting.com
Printed in the USA
LVOW02s0049110315

429969LV00004BA/6/P